Hippolyte Fontaine

Electrolysis

A practical treatise on nickeling, coppering, gilding, silvering, the refining of metals

and treatment of ores, by means of electricity

Hippolyte Fontaine

Electrolysis
A practical treatise on nickeling, coppering, gilding, silvering, the refining of metals and treatment of ores, by means of electricity

ISBN/EAN: 9783337406196

Printed in Europe, USA, Canada, Australia, Japan

Cover: Foto ©berggeist007 / pixelio.de

More available books at **www.hansebooks.com**

A PRACTICAL TREATISE ON

NICKELING, COPPERING, GILDING, SILVERING,

THE REFINING OF METALS AND TREATMENT OF ORES,

BY MEANS OF ELECTRICITY.

By HIPPOLYTE FONTAINE.

TRANSLATED FROM THE FRENCH BY

J. A. BERLY, C.E., A.S.T.E., &c.

34 ILLUSTRATIONS.

E. & F. N. SPON, 125, STRAND, LONDON.

NEW YORK: 35, MURRAY STREET.

1885.

PREFACE.

TREATISES on Electrolysis are fewer in France than in England, in the United States, and in Germany. This is not because the subject has been less studied in France than in other countries, but because French authors have published their works in the columns of scientific journals instead of publishing them in book shape.

Amongst the best known works, those of Sprague, Napier, Watt, Urquhart, Gore, and Wilson have passed through many editions in England; and those of Kafelowski, Scelhorft, Van Kress, Julius Weiss, Japing, Hans Jahn, and Pfanhauser are held in great estimation in Austria and Germany. Italy is beginning to enrich herself with some remarkable works upon the subject.

In France, only two practical treatises on Electrolysis are known which are of a real value:—'Guide du doreur, de l'argenteur et du galvanoplaste' (Guide for Gilding, Silvering, and Electro-depositing), by Roseleur; and 'Eléments d'électro-chimie' (Elements of Electro-chemistry), by Becquerel.

These two treatises complete each other, but are antiquated, having been written long before the recent progress made in electric generators, so that French technological literature, although so rich in theoretical and practical documents concerning almost every other branch of industry, shows in the speciality with which we are now dealing, a lamentable deficiency.

Our object in publishing this work is to make up for this deficiency to the best of our ability.

In order to facilitate the study of the subject, we have divided the work into four sections.

The first section is devoted to the theoretical notions indispensible to experimenters who wish to understand the electrical phenomena which they are daily witnessing; the second deals with the description of the principal batteries and dynamo machines used in electro-chemistry; the third section treats of electro-deposition: nickeling, silvering, gilding, coppering, &c.; and the fourth relates to the refining of metals and the treatment of ores.

Two of the chapters are of an exceptional length, those dealing with nickeling and the refining of copper. The relations which for the last twelve years we have had with a number of specialists, and our own private experience, have enabled us to collect a very precise and detailed series of information upon these subjects.

The theoretical part can easily be understood by practical men, consisting, as it does, of only very simple formulæ and elementary calculations, completed by a series of recapitulating tables on conductivity, electro-chemical equivalents, the heat generated by the formation of compound bodies, and the limits of electrolysis.

CONTENTS.

SECTION I.
THEORY.

CHAPTER I.
PRELIMINARY NOTIONS.

CHAPTER II.
ELECTROLYTIC LAWS.

CHAPTER III.
WORK ABSORBED IN ELECTROLYSIS.

CHAPTER IV.

TABULAR DATA.

SECTION II.

SOURCES OF ELECTRICITY.

CHAPTER V.

BATTERIES.

CHAPTER VI.

DYNAMO-ELECTRIC MACHINES.

CHAPTER VII.

CAPACITY AND EFFICIENCY OF ELECTRIC GENERATORS.

SECTION III.

ELECTRO-DEPOSITION.

CHAPTER VIII.

NICKEL-PLATING.

CHAPTER IX.

SILVER AND GOLD PLATING.

CHAPTER X.

COPPERING.

CHAPTER XI.

ELECTROPLATING AND ELECTROTYPING.

CHAPTER XII.

DEPOSITION OF VARIOUS METALS.

SECTION IV.

ELECTRO-METALLURGY.

CHAPTER XIII.

REFINING OF COPPER AND LEAD.

CHAPTER XIV.

TREATMENT OF ORES.

LIST OF ILLUSTRATIONS.

———•———

ELECTROLYSIS.

SECTION I.

THEORY.

CHAPTER I.

PRELIMINARY NOTIONS.

Magnetism—Energy—Hypotheses on Electric **Phenomena—Electric** Potential —Electromotive Force—Electric Current—Current Intensity—Electric Resistance—Electric Conductivity—Theoretical Electrical Units—Practical **Units—Ohm—Ampere—Volt—Dyne—Watt—Ohm's** Law—Electrical Work.

MAGNETISM.—Before entering upon the subject of the examination of the laws of electrolysis, it is necessary to briefly review certain notions on magnetism and on the principal phenomena accompanying the production of electricity.

The magnetic property of the natural substance oxide of iron, Fe_3O_4,—or loadstone—of attracting iron filings is very generally known, as also the fact that this property can be transmitted to bars of steel and artificially to soft iron. What is less generally known is that the magnetic and electric phenomena exert upon each other mutual reactions giving rise to the powerful electric currents used in industry.

The steel magnets, the only ones in use in the small laboratory apparatuses, are sometimes straight, or in the form of a horseshoe or of a half-circle, or, again, in the shape of a lozenge or of an arrow; sometimes they are made into cylindrical rings entirely closed, &c., &c.; but whatever their shape or dimensions, they all have certain regions, generally two, where their attracting property is most developed, and some intermediate parts where that property is nil. The regions of activity are generally at the extremities of the magnet, and are called poles; whereas the non-active part lies in the middle, and is called neutral zone or neutral line. When there exist more than two active regions, those at the extremities of the magnet retain the name of poles, and the others are called consequent poles. In a closed magnet the poles are situated at the extremities of a same diameter and the neutral zones are situated at the extremities of a diameter perpendicular to the latter.

When studying the magnets, Ampère came to the conclusion that each molecule of a magnetic body was a magnet in itself, having two poles and a particular direction, the whole being constituted in such a fashion that the total of the actions on any external point is nil. Under the influence of an external action the molecules set themselves in a special direction, in which they remain in steel and in all bodies possessed of a coercive force. In soft iron, the effects cease with the causes which produce them. According to Ampère's theory, the magnetisation of molecules is the result of permanent currents circulating round each of them.

When a steel needle is suspended by its centre, it is in indifferent equilibrium; but if it is magnetised, it will take up a determined position, to which it will revert whenever turned from it. This particular direction is the magnetic meridian, which differs very little from the terrestrial meridian.

The extremity of the needle pointing towards the north is called North Pole, whereas the other extremity, pointing towards the south of the earth, is called South Pole.

A magnetised bar exerts, in its neighbourhood, a very peculiar action; for example, if a suspended steel or iron needle is brought within a certain distance of it, the said needle will set in a well-determined position, according to its distance from the poles or from the neutral zone. The oscillations described by the needle before setting in an equilibrium position are indicative of the power of attraction in each point;

when they are short and quick, this power is intense; when they are long and slow, the power is feeble.

The magnetic field of a magnet is the space surrounding it, and submitted to or pervaded by its magnetic influence. As long as the needle above referred to is not in a state of equilibrium it is in the magnetic field of the magnet.

In order to study the magnetic properties of a magnet, the magnet may be placed under a sheet of cardboard or of glass, and iron filings spread from a sieve placed at a certain distance above it. In their fall the particles of iron arrange themselves according to well-defined directions, and in masses the more compact the nearer they are situated to the poles. The figures thus formed have been called magnetic phantoms. The curves formed by the particles of iron are called lines of force. Faraday has demonstrated that: (1) a line of force has always a tendency to be as short as possible; (2) that two parallel lines of force of the same intensity repel each other; (3) that the number of lines of force passing in each point is in proportion to the magnitude of the force in this particular point.

When two magnets are near each other, their lines of force exert on each other actions which, according to Faraday's law, alter their directions and respective intensities.

Coulomb has experimentally established that the magnetic attractions and repulsions exerted between two poles are inversely proportional to the square of the distance between them.[*]

ENERGY.—The energy of any given system is the total quantity of motive force which it possesses or can develop; in other words, it is the total amount of work which this system is capable of producing.[†]

The mechanical work resulting from the fall of a body; the calorific work due to variations of temperature; the chemical work which takes place in every transformation of substances; the electrical work obtained from the reciprocal motion of conductors, &c., are no other than special forms of energy. To

---

[*] The scope of this book does not allow us to enter more in detail into the analysis of magnetic phenomena. See for particulars 'Sur la théorie de la machine Gramme,' by A. Bréguet (Gauthier-Villars, 1880).

[†] É. E. Blavier, ' Des Grandeurs électriques.'

condense in a few words the definition of energy: it is the capability of effecting work. This definition equally applies to all natural forces, and even to those which are in a state of repose.

Energy is potential, or actual: potential when, in order to manifest itself, it requires a determinative cause; actual or kinetic when it is in action at the moment of being considered. The steam in a boiler, gunpowder, a coiled spring, a piece of coal, all contain a certain quantity of potential energy. The fall of a body, the steam acting in a cylinder, a current passing through a chemical battery, all develop actual or kinetic energy. There exists, in the universe, a total quantity of energy which is susceptible of infinite transformation, but the totality of which is an invariable quantity. It is quite as impossible to annihilate or create a quantity, however small, of energy, as to create or annihilate matter: all the phenomena of nature are in reality nothing but transformations of energy, and if it is sometimes difficult to explain what has become of some mechanical work which, as such, has entirely disappeared, one can be certain that it has been wholly transformed into heat, electricity, chemical work, &c.—in other words, into one or more forms of energy.

This principle, which is named conservation of energy or of motive forces, was explained in all its details, for the first time, by Dr. Helmholtz, in 1847, and it has contributed in a large measure to the progress of the mechanical, chemical, and electrical industries. When the experimenters have become perfectly convinced that energy of whatever kind could not disappear without instantaneously reappearing under some other forms, they have searched for the absent energy through all possible transformations without feeling discouraged by the difficulties—sometimes considerable—of investigation, and they have always succeeded in integrally finding, under the most varied and sometimes the most unsuspected forms, all that energy. We shall see, further on, that electricity, instead of being considered as a form of energy, can be considered as a simple transformator of energy. As M. Raynaud said with much reason: "Machines of all kinds are nothing more than tools for the distribution and repartition of energy; they restore, under the form of mechanical work or of heat, the energy which has been communicated to them in one or the other of these forms by natural forces or agencies, such as waterfalls, muscular power, air in motion, chemical actions, &c. The electricity developed in the course of this transformation is only an intermediary, and has no other function but to store energy, since it always restores it under a mechanical, calorific, or chemical form."

HYPOTHESIS ON ELECTRIC PHENOMENA.—In order to explain the electric phenomena, various theories have been put forward which, although not possessing the characteristics of absolute certainty, at all events enable one to realise the facts and to form a sufficiently approximate idea of the actions which are taking place. Amongst these theories, the two following are the more generally adopted.

Symmer admits the existence of two fluids, the one positive, the other negative, and entering into the composition of every substance. A body containing equal quantities of these two fluids would be in a neutral state; and a body containing unequal quantities of both fluids would be positively or negatively electrified according to the nature of the fluid in excess. That excess constitutes the free electricity of the body. The particles of a similar fluid repel each other and attract those of the other fluid. The power of attraction or of repulsion is in direct ratio to the square of the distance. Symmer says that the propagation of electricity in a conductor is the result of decompositions and recompositions of the neutral fluid.

Franklin admitted the existence of one fluid only, the molecules of which repel each other but attract ponderable matter. This is equivalent to the conception of a non-ponderable agent called electricity or electric fluid, existing in every body, in quantities varying according to the nature of the said bodies. If, as the result of a special action exerted upon any other body, it becomes possessed of more electricity than in its neutral state, the body is said to be positively electrified; if, on the contrary, the quantity of electricity which it contains diminishes, it is said to be negatively electrified. In this theory the phenomena of induction are explained by this consideration, that when an electrified body is in the vicinity of a neutral body, there is an action of the former upon the latter: the forces resulting therefrom disturb the equilibrium of the neutral body and cause a new distribution of electricity to take place.

Electrified bodies are possessed of the property of attracting light bodies; in other words, when a metallic wire is in an electric circuit, it possesses magnetic properties easy to verify. If iron filings are projected around a vertical conductor through which an electric current is passing, a phantom, analogous to that of magnets, will be formed. The lines of force are closed curves following the same laws as those produced by magnets.

If two currents flow near each other in a same direction their lines of force are of elliptic shape, the wires being at the focuses; should the two currents flow in opposite directions, the lines of force form two clearly distinct systems.

Ampère has demonstrated that two parallel currents flowing in the same direction attract, and two parallel currents flowing in opposite directions repel each other.

These laws appear at first sight to be contrary to the law of Coulomb, but it should be observed that Ampère treated of currents, and that Coulomb dealt especially with electrostatics and electric particles. The following is Coulomb's law :—Two electric particles exert upon each other a repulsive or an attractive action, according as they are of the same or of contrary names ; the said action varies proportionally to the masses of the acting bodies, and in an inverse ratio to the square of their distance.

ELECTRIC POTENTIAL.—As we have already said, nothing certain on the exact nature of electricity is known, and in the actual state of science, the existence of fluids carrying with them the idea of action at distance is only explained by some very doubtful hypotheses. This mystery, which surrounds the causes of the electric phenomena, is not, however, peculiar to manifestations of the current only ; nothing is known of the nature of light, heat, motion, &c. All the forms of energy can be verified, measured, modified, or entirely transformed, but it has as yet been found impossible to discover their initial causes. The want of any rational explanation of the exact origin of facts causes us to remain in the field of hypotheses, and to designate the initial causes by certain names without being able to define them with absolute accuracy. It is, therefore, important that the new designations should be made well known by comparing them with the physical, mechanical, and chemical phenomena which are more generally understood.

For instance, the special **property of** electricity **which** corresponds to the expanding force of **gases, to the hydrostatic** pressure of liquids, to the temperature of bodies **in** thermodynamics, is called electric potential. **It is owing to the** difference of potential in two points, that electricity **will travel** from one to the other. Two bodies are said **to have the same** potential when, on being connected by a conductor, no electric motion takes place.*

A difference **of** potential, like a difference of level, is indicative of a condition of relation between two points without involving the necessity of ascertaining the exact value of the potential at those points. **In** the same manner as hydraulic work is calculated by multiplying **the** weight of the water by the head of the fall, so is electrical work calculated by multiplying the quantity of **electricity in motion by** the difference of potential.

Two conducting bodies joined by a metallic wire become one single conductor and set at the same potential. The earth **may** be considered as an immense conductor the potential of which is nil : an electrified body therefore sets at a nil potential when it is put in communication with the ground.

The electric capacity of an insulated conductor is the quantity of electricity which is required to raise its potential by one unit.

The electric quantity is measured by the power of attraction developed by the current. If a quantity A of electricity exerts a force double that of a quantity B, under the same conditions, **the** quantity A is said to be double the quantity B.

ELECTROMOTIVE FORCE.—The electromotive force is that particular **force, whatever** its origin, which tends to produce **a** displacement **of electricity.** The difference of potential between two points gives the measure of the electromotive force.

---

* 'Des Grandeurs électriques,' by **E. E. Blavier.**

It is well to observe that there can be a manifestation of electromotive force in a circuit which is brought towards or away from another circuit without any difference of potential. It is, according to Ampère, thus that the currents produced by terrestrial magnetism should be considered. Electromotive force, notwithstanding that it can be measured, and can be produced by a difference of potential, has therefore a more general signification.

ELECTRIC CURRENT.—Whenever a metallic communication is established between two insulated conductors at different potentials, the equilibrium cannot subsist, the positive electricity passes from the body at the higher potential towards the other, a flow of electricity being produced which is designated by the name of electric current. If the two bodies contain limited charges, the state of equilibrium is rapidly reached, and the current obtained varies with the time. But if by any means the difference of potential between the two conductors is maintained constant, a permanent regimen is established, and the electric current is itself constant and regular.

CURRENT INTENSITY.—Electric currents are possessed of different properties; some, as the heating of conductors, chemical decompositions, magnetic induction, require a transformation of energy; some others, as the action of one current upon another current * or upon a magnetic pole, can be observed in a static condition, that is to say without any production of work, by the simple measurement of a force.

By definition the intensity of a current in a conductor is proportional to the quantity of current flowing through the section of the said conductor in a unit of time. When a state of stability has been established, or, otherwise, when the flow of electricity is regular and constant, the intensity is the same at every point of the circuit.

Faraday and Pouillet have proved experimentally that the intensity of a current is proportional to its action upon the pole of a magnet or upon another constant current, which affords an easy means of measuring it.

ELECTRIC RESISTANCE.—All conductors offer a resistance to

---

* 'Leçons sur l'électricité et le magnétisme,' par Mascart et Joubert.

the passage of **the current.** Whatever may be the conductor connecting **two bodies** charged with electricity at different potentials, a certain time must lapse before the equilibrium **is** established. The quantity of electricity flowing in one second between two points, the electromotive force of the current being maintained constant, depends on the resistance of the conductor connecting these two points. The same principle **can be** expressed under the following form : The **resistance of conductors** is proportional to the speed of the **current.**

From the foregoing it will **be seen that the electric resistance** is the property possessed **by conductors to cause variations in** the intensity of a current **produced by a** given **electromotive** force. The resistance **of a** body is directly proportional to **its** length and inversely **to** its section and to its conductivity.[*]

The specific resistance of a substance is the electric resistance of that substance estimated in relation **to** the units **of** length and of volume ; **it is** that which would **be** offered by a prismatic conductor, **made of** the **said substance, and** the **section** of which would be equal **to the unit of surface.** Adopting the centimetre as the unit **of length it would be** that of a cubic centimetre, **two** opposite **surfaces of** which **were** maintained **at** constant **potentials. The** specific resistance can also **be estimated in relation to the unit of** mass ; it is then the **resistance** of a conductor, **the mass of which** weighs one **gramme, and of a** section of one **square centimetre.**

The resistance is frequently expressed, in theoretical **calculations,** by the relation **of a** time to a length, that **is to say, by** the converse of a speed.

CONDUCTIVITY.—Conductivity is, **as its name** indicates, the property possessed **by** substances to **act as** vehicles of electricity ; it is the converse **of** resistance. **Calling** R **the** resistance, the conductivity will **be** represented by $\frac{1}{R}$, and, since the resistance **corresponds to** the converse of a speed, the conductivity may be expressed **in a** speed **or by** the relation of a length to a time.

THEORETICAL ELECTRICAL UNITS.—In order to **compare** electric currents, units must be **used.** The number **of these**

* 'Des Grandeurs électriques,' **par E. E.** Blavier.

units must be sufficient to allow of the valuation of all the quantities which characterise electric currents, and also to estimate in dynamical work the energy to which they correspond.

Physical phenomena being always dependent on the three elements: matter, space, time, it is necessary to have a unit of mass for estimating the quantity of matter, a unit of length for comparing the spaces, and a unit of time for estimating the duration of a phenomenon.

These three units, which are called fundamental, and have been finally determined at the International Congress of 1881, are:—

Unit of length :	L = 1 centimetre.
Unit of time :	T = 1 second.
Unit of matter :	M = the mass of 1 gramme.*

This system is known as the C. G. S. system (centimetre, gramme-mass, second).

The derived units, always in view of theoretical calculations, are:

Resistance †	.. $R = L\ T^{-1}$
Electromotive force	.. $E = M^{\frac{1}{2}} L^{\frac{3}{2}} T^{-1}$
Intensity	.. $C = M^{\frac{1}{2}} L^{\frac{1}{2}} T^{-1}$
Quantity	.. $Q = M^{\frac{1}{2}} L^{\frac{1}{2}}$
Capacity	.. $K = L^{-1} T^{2}$
Force	.. $F = ML\ T^{-2}$
Work	.. $W = ML^{2}\ T^{-2}$

---

* The decision of the Congress relating to the unit intended for estimating the value of matter has been very much criticised. The two other fundamental units having been selected amongst the usual practical systems, it was deemed advisable to select the third one from amongst the same. This amounts to saying that, by adopting the weight of 1 gramme instead of its mass, not only would the electrical calculations have been facilitated, but especially their comparison with mechanical calculations.

The Congress considered that the action of gravity varied with the altitude and latitude, so that one standard gramme carried from Paris to London, Berlin, or even to Versailles, would not exactly correspond to the weight of one cubic centimetre of water, and that a particular standard of size would be required at every point of the globe. The unit of mass is, on the contrary, invariable. The weight varies, but the acceleration which the gravity impresses upon it varies in the same ratio, so that $\dfrac{W\ (weight)}{g\ (gravity)}$ or the mass is invariable. This is the reason why the Congress, ratifying Gauss's notions, adopted, as a third fundamental unit, the mass of 1 gramme and not that of $9 \cdot 81$ grammes.

† We are using here the most usual formulæ, without positively insisting on

PRACTICAL UNITS.—The practical units, as decided upon by the Congress of 1881, and the Conference of 1884, are as follows : the ohm, the volt, the ampere, the farad, and the coulomb.

OHM.—The ohm is the practical unit of resistance ; its legal value is represented by a column of mercury of a section of one square millimetre, and 1·06 metre long, at the temperature of melting ice. This value corresponds, as near as it is possible to estimate it, to 1 milliard of theoretical units, that is to say, 1 milliard of centimetres divided by one second $\left(\dfrac{\text{the quarter of the terrestrial meridian}}{\text{one second}}\right)$.

AMPERE.—The ampere is the practical unit of intensity; its value is equal to $\frac{1}{10}$ of the theoretical unit C.G.S. For industrial purposes the ampere is generally taken as being representative of the quantity of silver deposited per second. According to the recent calculations of Messrs. F. and W. Kohlrausch, this quantity is equal to 0·00111888 gramme.

VOLT.—The volt is the practical unit of electromotive force ; it is the electromotive force which sustains a current of one ampere through a resistance equal to the legal ohm.

The material value of one volt is approximately equal to the electromotive force of one zinc-copper cell with dilute sulphuric acid and sulphate of copper, known as the Daniell cell.

COULOMB.—The coulomb is the unit of electric quantity. It is the quantity of electricity passing through a circuit during one second when the current is equal to one ampere.

FARAD.—The farad is the unit of capacity. It is defined by the condition that one coulomb in one farad gives one volt.

The British Association has adopted the dyne as a unit of force, and the erg as a unit of work. We will describe them, although they are not used in practice.

DYNE.—The dyne is the force which, acting upon the unit of mass (1 gramme), impels it at a speed of one centimetre at the

---

these units, which are not used for industrial purposes. The formula of resistance, $R = L\,T^{-1}$, may be written $\dfrac{L}{T} = \dfrac{1\ \text{centimetre}}{1\ \text{second}}$; the formula of force, $F = M\,L\,T^{-1}$, may be written $\dfrac{M\,L}{T^2}$, and be expressed as the mass of 1 gramme multiplied by 1 centimetre and divided by the square of a second.

end of the first second. A body, falling in vacuum, acquiring a speed of 981 centimetres at the end of the first second, the value of a dyne will be $\dfrac{1 \text{ gramme}}{g} = \dfrac{1}{981}$ gramme, or about one milligramme.

ERG.—The erg is the unit of work. It is the product of the unit of force by the unit of length, its value being, therefore, $\dfrac{1}{981} \times 0 \cdot 01^{m} = \dfrac{1}{98,100}$ grammetre, or $\dfrac{1}{98,100,000}$ kilogrammetre. One kilogrammetre is, consequently, equal to about 100 millions of ergs. From the foregoing, it is easy to establish the relations existing between the mechanical and electrical units. The mechanical equivalent of heat is 425, which means that one calorie, or the quantity of heat necessary for raising one degree centigrade one litre of water, is equal to 425 kilogrammetres. One erg will consequently be equal to $\dfrac{1}{98,100,000 \times 425}$ calories.

One calorie, therefore, is equivalent to 41,692,500,000 ergs.

As will be seen, the dyne and the erg are extremely small quantities which cannot be utilised for industrial purposes. The kilogramme and kilogrammetre are, on the contrary, of ready use, whether in connection with mechanical, chemical, calorific, or electrical energy.

WATT.—The late Sir William Siemens proposed to call watt the product of one volt by one ampere; this renders possible the measuring the rate of production of an electric machine. Wires of varying diameters can be wound on the frame of a dynamo machine. When the wire is thin, the electromotive force obtained is great, and the intensity small; when, on the contrary, the wire is thick, the electromotive force is small and the intensity great; but in all the combinations, the product of the volts by the amperes is sensibly the same. The watt may therefore be used as a comparison between various forms and sizes of electric machines.

OHM'S LAW.—The law of Ohm gives the relation which exists between the electromotive force, the intensity, and the resistance of a current. Calling E the electromotive force;

C the intensity; R the resistance; C is equal to $\frac{E}{R}$, from which

$E = CR$ and $R = \frac{E}{C}$ or, in other terms: the intensity of a current is obtained by dividing the electromotive force by the resistance; the electromotive force is equal to the product of the intensity by the resistance; and the resistance is equal to the quotient of the electromotive force divided by the intensity. These formulæ are of common use in laboratories and in the electrical industries.

ELECTRICAL WORK.—The quantity of energy of a current is equal, in kilogrammetres, to the product of the amperes multiplied by the volts and divided by g (9·81).

$$W = \frac{CE}{g} = \frac{CE}{9 \cdot 81} \text{ kilogrammetres.}$$

By combining this equation with those derived from Ohm's law, the work in relation to resistance can be expressed as follows:

$$W = \frac{E^2}{g R} \quad \text{and} \quad W = \frac{R C^2}{g}.$$

Thus the electrical energy or work which can be obtained from a given current can be calculated by either multiplying the ohms by the square of the amperes and dividing by 9·81; or by multiplying the amperes by the volts, and dividing by 9·81; or again, by dividing the square of the volts by the product of the ohms multiplied by 9·81.

A sufficient approximation is obtained by substituting 10 to 9·81; this simplifies the calculations, which, however, do not present any practical difficulties.

For example, if a machine gives a current of 750 volts and 10 amperes, it can be at once asserted that it develops a work equal to $\frac{750 \times 10}{10} = 750$ kilogrammetres, or 10 horse-power; and should the current be sent through another dynamo having a known efficiency of 80 per 100, Prony's brake indicates a work of 8 horse-power really produced on the spindle of that second machine.

CHAPTER II.

ELECTROLYTIC LAWS.

Chemical Action and Electricity—Effects of Electricity—Electrolysis—Clausius'
Hypothesis—Electrolytic **Action** of Currents—Faraday's Laws—Electro-
chemical Equivalents—Of the Influence of Solutions—Of the Influence of
the Sizes **of the** Electrodes—Joule's Law—Work required in Electrolysis
—Electromotive Forces—Thomson's **Law**—Determination of the Electro-
motive Force—Electromotive Force **required for the** Decomposition of
Water—Remark on the **Decomposition of Water by means of a feeble**
Electromotive Force.

CHEMICAL ACTION AND ELECTRICITY.—Whenever a **chemical**
bath liable to be decomposed **is traversed by an electric current,**
the latter causes the separation **of its parts ; and, reciprocally,**
each time the molecular equilibrium **of a body is disturbed by**
some chemical affinities, there is a production of **electricity.**

In a combination, **the basic body** becomes charged with
negative and the **acid body with positive** electricity. In a
decomposition, **the reverse takes place, the** basic body taking
the positive, **and the acid body the negative** electricity.

EFFECTS OF ELECTRICITY.—**Electricity** acts by discharges
or by **current.**

When acting **by discharges, it gives rise to** gaseous **com-**
binations **and** decompositions, **or to** modifications in the properties of simple **bodies; for instance,** when pure oxygen is
submitted to electric effluvia, it is reduced in volume and
acquires that peculiar odour which has been called ozone.

When, acting **under the form of a current,** electricity
passes through **certain liquids, it will decompose** them into
their constituents; **the metals, the bases,** and the hydrogen
flowing **to** the negative pole, **and the acids** and oxygen being
liberated at the positive pole.

Faraday discovered **and also formulated** the principal laws
regulating the chemical **actions of electric** currents; to that

illustrious *savant* is also due the knowledge of the induction currents now in general use in the industrial operations of electroplating.

ELECTROLYSIS.—Faraday has given the name of electrolysis to the process of decomposition by means of the electric current. He termed *electrolyte* the liquid in which the decomposition occurs; *electrodes*, the conductors immersed in the liquid ; *anode,* the electrode corresponding to the positive pole of the electrical source, and *cathode* the negative electrode. Faraday has termed *ions* the substances produced by electrolysis and which are divided into *anions* and *cations*, according to their appearing on the anode or the cathode. These three last terms are very little used, but the others are part of the scientific language of every country.

Two conditions are indispensable in order to produce electrolysis : the substance which is to be decomposed must be a conductor of electricity, and in a liquid state. The said substance is generally dissolved in water ; but water is not absolutely necessary to electrolysis as a few specialists will have it ; chloride of magnesium, for instance, can be electrolysed after it has been brought to a liquid state by means of an igneous fusion; the essential point is to obtain a liquid, no matter the means by which the liquefaction is produced.

CLAUSIUS' HYPOTHESIS.—The real cause of the chemical action of a current is no more known than the initial cause of the current itself; the theory of electrolysis, therefore, is also dependent on mere suppositions. But as, owing to Faraday's discoveries, the laws regulating electro-chemical work are known, the hypotheses, however ingenious they may be, concerning the original causes of the said work, are of no more than secondary importance. We will therefore only summarily mention the most universally accepted hypothesis, not with a view of confirming the formulæ which will be given hereafter, but in order simply to give to practisers a general explanation of the facts of which they are daily witnesses.

Clausius admits that matter is constituted of extremely small particles called molecules, which cannot mechanically but can chemically be divided ; the molecules of simple bodies are identical, those of compound bodies differ. The constituent parts of molecules which can neither be divided mechanically nor chemically are termed atoms. The molecules of bodies are animated by a continual motion, and are possessed of an initial force which is constantly dependent on their temperature. In solid bodies, the molecules are only possessed of a vibratory motion, and cannot abandon their position of equilibrium. In liquids, on the contrary, although appearing motionless, these molecules turn, roll over, and knock against each other, travelling from one point to another without being subjected to return to their original position. In a gaseous state the molecules are knocking against each other, whirling, striking the walls of the vessel inclosing them, and thereby find themselves outside the spheres of mutual attraction.

The elementary atoms of compound bodies are some electropositive and the others electro-negative; they are not invariably united in liquids, and each passes from one molecule into another, setting at liberty another atom similar to itself, and which, in its turn, decomposes another molecule. As soon as a current passes through a compound liquid, the molecules which previously had arbitrary motions in every direction, begin to shift in a regular manner. The electro-negative atoms are directed towards the positive electrode, and the electro-positive atoms towards the negative electrode.

Thus, after Clausius, the atom of hydrogen which constitutes a part of a molecule of water is not invariably connected to the corresponding atom of oxygen; but is carried away into an unceasing whirling, can part from this particular atom of oxygen in order to combine with a neighbouring atom of oxygen and get so carried, by means of successive exchanges, to distances proportionally indefinite considering the radius of activity in which the phenomenon is taking place. The effect of the current of electricity is to impress to these motions—to these exchanges —a systematic tendency in virtue of which the atoms of hydrogen are directed towards the negative electrode, whereas the atoms of oxygen travel towards the positive electrode.[*]

Before this hypothesis was formulated, Faraday had

* See the works of Messrs. Mascart and Joubert, Blavier, Gariel.

remarked that an electric current could pass through water, but not through ice, although the composition of the two substances is identical; he concluded from this that in the liquid state the molecules were allowed to place **themselves in** the direction of the line of polarisation, whereas **the** rigidity **of the solid** state was opposed to such a direction of the molecules taking place.

ELECTROLYTIC ACTION OF CURRENTS.—In all chemical baths which **are** being electrically treated, the current **enters** through the anode, travels across the liquid space comprised **between** the anode and the cathode, and leaves through **the cathode.**

It is during the space **of time taken by the current to travel** between the two electrodes that the action **of** electrolysis takes place.

If, in order to fix ideas, we **assume that** an electric current **is** passing through a binary **compound,** liquid and conductive, as, for example, **chloride of** copper, **the** current decomposes the chloride into **its two** constituents; **the** chlorine going to the positive pole or anode, and the pure copper depositing itself on the cathode. This decomposition requires a certain expenditure of work, which we will analyse further on.

The metal always goes **to the negative, and** the other constituents to the positive **pole. If** operating on **a sulphate of** copper **bath, the** copper **will be** deposited **on** the **cathode or** negative pole, **and** both the oxygen and the sulphuric **acid will** go **to** the anode **or** positive pole. **There,** however, sometimes occur some secondary actions which **cause the** primordial effect to be disturbed.

FARADAY'S LAWS.—*First law:* **The** quantity **of substance** decomposed in a given time is proportional to the intensity of the current, or, in other words, to the **quantity of** electricity **passing** through the liquid. This **law may be** verified by **means of** a water voltameter, the discharges **of a** condenser, or the current of a battery, or of an induction machine being used for the purpose. The quantity of **gas** generated in each of the tubes of **the voltameter—for** a quantity of electricity equal **to** two, **three, four, &c., fold that** of a first experiment—will always be corresponding **to** the quantity of current passed through **it.**

A very important fact results from this first law, **viz.** that in order to decompose one equivalent of any given substance, for instance, one gramme of hydrogen, a constant quantity of electricity is required, which may be called electro-chemical equivalents.

Second law.—When the same current simultaneously acts on a series of solutions, the weights of the constituents separated in each of them are in the same ratio as their chemical equivalents. Supposing, for instance, that the current is made to traverse three vats containing, the first some water, the second some nitrate of copper, and the third some nitrate of silver; for each gramme of hydrogen set free in the first vat will be found, in the second bath 31·75 grammes of copper, and in the third 108 grammes of silver; the figures 1, 31·75, and 108 being respectively the equivalents of hydrogen, copper, and silver. This second law clearly shows that the electric intensity is the same in every point of the circuit whatever may be the nature and the number of successive conductors of which the said circuit is composed. If a series of water volta-meters of widely varying dimensions are placed in the same circuit, it is found that the quantities of hydrogen and oxygen liberated are the same in all the voltameters, although the degree of acidity and of conductibility of the water might have varied in each voltameter.*

Third law.—The electrolytic action is independent of the respective position of the battery and of the electrolyte. This law is easy of demonstration, as it will be found that wherever the battery is placed, that is to say, whether in the bath itself or at a great distance from it, the quantity of electrolyte decomposed does not vary, the conditions of the current remaining, of course, the same.

Fourth law.—The number of equivalents of zinc dissolved in each cell of the battery, is equal to the number of equivalents of metal liberated in each electrolyte which is part of the

---

* In the application of the voltameter for the measure of feeble currents it must be observed that the quantities of gas liberated in one apparatus, owing to the decomposition of water, do not depend only on the initial intensity of the current, but also on the degree of acidity of the water, the nature and dimensions of electrodes, the distance between the anode and cathode; these various causes creating some resistances of a greater or lesser magnitude, and reducing the initial intensity of the current.

circuit. This law, which is **expressed as** regards a zinc **cell** used as a **generator** of current, is general, and is verified **with** any other **cell**. It **is** a consequence of the second law which demonstrates **that the** current **acts in** the electrolytes in the same manner as in the cell, and that the chemical work of **a** current is the same in the whole of its course.

All the foregoing can be verified by means of simple decompositions, provided a secondary action does **not come** into play; but, in practice, there often occur some secondary **phe-**nomena due to the chemical action of the constituents which are carried to the **electrodes, owing** either to **the decomposition** of the water of the bath by the metal in its natural **state, or to** the anode being attacked by the liberated acid and oxygen.

The effects of the electro-chemical decomposition in definite proportions, **vary** according to the temperature and the nature of the dissolving electrodes, and of the dissolved compound.

If we fail to verify Faraday's law, **it** would be owing to neglecting to take into account **one or** many of the causes which may influence the **result of electrolysis.**

ELECTRO-CHEMICAL EQUIVALENTS.—The electro-chemical equivalents are the weights of **various** electrolytes **decomposed** for each unit **of electricity.** The electro-chemical equivalents are proportional to their ordinary chemical equivalents.

There is no difficulty in defining the electro-chemical equiva-**lents** of analogous chemical **compounds; but** if **in the same** circuit the electrolysis **of water is** produced together **with that** of a series of **neutral sulphates of** protoxide, for **example, the** electro-chemical equivalent of each metal is the weight which is deposited for the liberating of $0 \cdot 00001036$ gramme of hydrogen. If the compound substances **have** not **the** same formula, one may experience **some** difficulty with **two** neutral sulphates, one of protoxide, **the other of** sesquioxide of iron decomposed by the same **current;** it **may** be asked if the same **weight of iron or** of oxygen **will be liberated in the** two baths. M. Ed. Becquerel has demonstrated **that it** is the metalloid which **rules the case.** Consequently, the weights of iron, in the **two** electrolytes, **will** be in the ratio **of** 3 to 2. The same **rule** applies to the salts of other acids, the chlorides, sulphides, &c.[*]

* ' Leçons sur l'électricité et le magnétisme,' par E. Mascart et Joubert.

The law regulating electro-chemical decomposition can, according to M. Ed. Becquerel, be formulated as follows, facts being generalised: When an electric current *passes* through two or more binary compounds, the decomposition always occurs in definite proportions, so that, for each equivalent of electricity a chemical equivalent of the substance acting as an acid, or otherwise of the electro-negative constituent, goes to the positive pole, a corresponding quantity of the electro-positive element being deposited at the negative pole. M. Ed. Becquerel has also deduced from his researches the following rule: If one equivalent of a substance, either simple or compound, combines with one or more equivalents of another substance, and the first acts as an acid or electro-negative element in the combination, the production of electricity resulting from their chemical action represents the equivalent of electricity. The conclusion is that the quantity of electricity brought into action solely depends on the substance which acts as an acid.

INFLUENCE OF SOLUTIONS.—Faraday's laws are, in practice, verified with such slight discrepancies that these can always be attributed to inaccuracy of calculations, or to experimental errors. M. Soret, who made the study of a large number of solutions, has always found that they were without any influence on the weight of the precipitated metal, as is proved by the few following examples, which are all reduced to the deposition effected by one ampere traversing during one hour a saturated solution of sulphate of copper.

Solutions.	Weight of Precipitated Copper in Grammes.	Difference with the Weight obtained with the Saturated Sulphate.
		grammes.
Saturated sulphate of copper	1·1812	..
Sulphate of copper diluted in 1 volume of water	1·1835	+ 0·0023
Concentrated nitrate of copper	1·1805	− 0·0007
Phosphate of copper dissolved in phosphoric acid	1·1800	− 0·0012
Acetate of copper	1·1797	− 0·0015
Mixture of sulphate of copper and sulphate of potassium	1·1806	− 0·0006
Mixture of sulphate of copper and acetate of cobalt	1·1825	+ 0·0013
Mixture of sulphate of copper and sulphate of zinc	1·1835	+ 0·0023
Mixture of sulphate of copper and sulphate of cadmium	1·1835	+ 0·0023

INFLUENCE OF THE SIZES OF THE ELECTRODES.—When electrolysis is carried on with electrodes of greatly different sizes, the results do not appear, at first sight, to be in accordance with the theory, but the apparent discrepancy is readily explained by the intervention of secondary actions. For instance, if the electric current, with a wire and a strip of platinum as electrodes, passes through a bath of dilute sulphuric acid, the following quantities of gases will be liberated, and these are not always in the ratio to the chemical equivalents :—

Negative strip.. .. .. 100 cubic centimetres	of hydrogen.	
Positive wire .. .. .. 50 "	"	oxygen.
Negative wire.. .. .. 41 "	"	hydrogen.
Positive strip .. .. .. 16 "	"	oxygen.

The oxygen missing in this last experiment has simply been absorbed by the platinum strip. If instead of platinum a strip of spongy platinum is used, the voltameter is found to contain only five volumes of oxygen against twenty of hydrogen liberated by the negative wire.

JOULE'S LAW.—An electric current, passing through a conductor, generates a certain quantity of heat. In the term conductor we include the generator in which the current is produced, as well as the apparatuses in which the work of electrolysis is performed, and the metallic rods establishing a continuous communication between every part of the circuit.

Joule has made a complete study of this phenomenon, and has established the following law :—

The quantity of heat developed in a conductor is proportional to the resistance of that conductor, and to the square of the intensity of the current.

If H is the total quantity of heat developed ;

t the time during which the current passes ;

C the intensity of the current ;

R the resistance of the conductor ;

A the mechanical equivalent of heat ($\dfrac{424}{9 \cdot 81}$ for the electrical units in which the mass and not the weight is concerned), we have—

$$H = \frac{C^2 R t}{A} \text{ calories.}$$

Numerous and precise experiments have established that
this law could in every case be verified; whether the heat
be in a sensible form in a calorimeter, or in the form of chemical
energy in an electrolyte.

Combined with Ohm's law, Joule's law can be expressed in
the two following formulæ:— · .

$$H = \frac{E\,C\,t}{A} \quad \text{and} \quad H = \frac{E^2\,t}{A\,R},$$

E being the electromotive force of the current.

We have already seen that the work due to a current is
equal to the square of the amperes multiplied by the resistance
and divided by 9·81.

Estimating, by means of Joule's law, the work absorbed by
a given resistance, the same formula will naturally be arrived
at; let us take first the general equation:

$$H = \frac{C^2\,R\,t}{A} \text{ calories.}$$

Work being equal to the number of calories multiplied by
the mechanical equivalent of heat,

$$W = HA = C^2\,R\,t.$$

Expressing the intensities in amperes, the electromotive
forces in volts, and the resistances in ohms, units which are
based on the mass and not on the weights, the formula becomes

$$W = \frac{C^2\,R}{9\cdot81} \text{ kilogrammetres per second.}$$

WORK REQUIRED IN ELECTROLYSIS.—The dynamic work
required for decomposing a given solution is equal to at least
that which corresponds to the heat produced by the decomposed
substances when recombining together in order to reconstitute
the original solution. This law, which is based on the principle
of the conservation of energy, is often difficult to verify, owing
to the complex secondary actions which escape the notice of
even the most clever experimenters; but it is indisputable and
must serve as a basis in every calculation relating to electro-
chemistry.

If it were possible to decompose substances which on recomposing produced more work than was required for their decomposition, the result would be the creation of energy, which is quite as impracticable as the realisation of perpetual motion.

ELECTROMOTIVE FORCES.—Faraday's law respecting the quantities of substances liberated by one unit of intensity, and the preceding one as regards the work required for effecting a chemical decomposition, establish that, to electrolyse a given compound it is necessary to use a given electromotive force, and whatever may be the intensity of the current, no decomposition of the electrolyte occurs if that electromotive force is not reached.

Calling E the electromotive force necessary for the decomposition of the bath, and Q the number of coulombs flowing through the bath, the work of decomposition will be expressed by the formula:

$$W = \frac{Q\,E}{9\cdot81} \text{ kilogrammetres,}$$

whence

$$E = \frac{W \times 9\cdot81}{Q}.$$

If z is the electro-chemical equivalent of the liberated substance, the total weight liberated by Q coulombs will be equal to $Q\,z$. Calling H the number of calories emitted by one gramme of the substance liberated by the electrolysis for returning to the state of combination it was in at the beginning of the operation, the heat produced by $Q\,z$ will be $Q\,z\,H$, and, as the mechanical equivalent of heat is $0\cdot424$ kilogrammetres for one calory (gramme-degree), the corresponding work will therefore be

$$W = 0\cdot424\,Q\,z\,H \text{ kilogrammetres.}$$

As we have already shown that

$$W = \frac{Q\,E}{9\cdot81},$$

after simplifying we will have

$$E = 4\cdot15944\,z\,H \text{ volts.}$$

THOMSON'S LAW.—Thomson's law is but a paraphrase of the above formula; it can be thus formulated: The electromotive force of an electrolyte is, in absolute measure, equal to the mechanical equivalent of the chemical action to which an electro-chemical equivalent of the decomposed metal is subject.

DETERMINATION OF THE ELECTROMOTIVE FORCE.—Observing that z H represents the heat given off by one gramme of the substance under consideration, multiplied by the electro-chemical equivalent of the said substance, the electromotive force can be estimated in respect of the chemical equivalents, and instead of E = $4 \cdot 15944\, z$ H volts, we can write E = $4 \cdot 16 \times 0 \cdot 0105$ H e = $0 \cdot 0434$ H e volts, e being the chemical equivalent of the metal acted upon, H e is the number of calories (kilogramme-degree) evolved by the combination of one chemical equivalent of the substance under consideration:

$$1 \text{ volt therefore corresponds to } \frac{1}{0 \cdot 0434} = 23 \text{ calories.}$$

In order to determine the electromotive force, in volts, necessary for chemical decompositions, it will therefore be sufficient to know the quantity of calories evolved by a chemical equivalent of the metal decomposed in the electrolytic bath and to divide that quantity by 23.

ELECTROMOTIVE FORCE REQUIRED FOR THE DECOMPOSITION OF WATER.—Let us take as an example the decomposition of water, in which hydrogen acts as the attacked metal. One gramme of hydrogen in being oxidised, can develop $34 \cdot 5$ calories, therefore

$$E = \frac{34 \cdot 5}{23} = 1 \cdot 495 \text{; about } 1\tfrac{1}{2} \text{ volt.}$$

REMARK ON THE DECOMPOSITION OF WATER BY MEANS OF A LOW ELECTROMOTIVE FORCE.—It will be seen from the foregoing that water cannot be electrically decomposed unless a current of at least $1 \cdot 495$ volts is used. Many physicists, however, maintain that this decomposition can be effected with a lower electromotive force, for example, by means of one Daniell cell, the electromotive force of which is approximately equal to one volt. Doctor Jahn himself, in his remarkable work on the theory of electrolysis,[*] explicitly says :—

" We have already established that the electromotive force of any circuit must be proportional to the production of heat

* ' Die Elecktrolyse und ihre Bedeutung für die theoretische und angewandte Chemie.' Vienna, 1883.

which corresponds to it in the chemical reactions. We **found,** for example, that the electromotive **force of** a Daniell cell is proportional **to the heat which** corresponds **to** the substitution of one equivalent of copper **to** one equivalent of zinc, and that consequently **it is** relatively measured by 25 calories, whereas the equivalent of hydrogen combining with **the** equivalent of oxygen produces $34 \cdot 5$ calories.

" There results, from the foregoing, **the fact** that **a** Daniell cell is not capable of decomposing one equivalent **of water in the** same time that one equivalent **of** zinc **is** dissolved **in the cell.**"

Dr. Jahn adds :

" As according **to an** acknowledged **principle in** thermo-chemistry the decomposition **of** one combination absorbs exactly **the** same quantity **of heat** as evolved **by its** formation, the de-composition **of 9** grammes of water **requires about** $34 \cdot 5$ calories, that is to say, a quantity of heat **which** a Daniell cell can in no case furnish by the dissolution **of one** equivalent of zinc. I par-ticularly insist on **this** latter restriction as it has been a frequent custom to give this proposition the appearance of an impossibility for a Daniell cell to decompose **water.** But this is contradicted by experience."

Of all the theories and laws applied to electrolysis, Dr. Jahn only accepts those of Faraday, **that** of the conservation of energy and that of the **electro-chemical** equivalence. We are exactly of his opinion, **and it is preciscly** because we accept these laws that we dispute his reasoning.

If we admit, for one moment, that it is possible to decompose water with an electromotive force of one volt, as, according to Faraday, one ampere will always liberate the same quantity of water, viz. $0 \cdot 00000009328$ kilogramme per second, the work necessary for the decomposition of **one** kilogramme of water will be :

$$W = \frac{1}{9 \cdot 81 \times 0 \cdot 00000009328} = 1092800 \text{ kilogrammetres.}$$

On **the other hand the** combustion of the hydrogen contained in one kilogramme of water corresponds to

$$111 \times 34 \cdot 5 \times 424 = 1623708 \text{ kilogrammetres.}$$

By expending a work of 1092800 kilogrammetres it would therefore be possible to obtain a chemical energy capable of producing 1623708 kilogrammetres ; this would involve the creation of a certain quantity of energy, which is impossible.

If we admit that the quantity of decomposed water is proportional to the work of dissociation, we find that to equilibrate the work of recomposition equal to the electro-chemical equivalent of water

$$x = \frac{1}{9 \cdot 81 \times 1623708} = 0 \cdot 0000000627 \text{ kilogramme.}$$

So that one ampere, which ought to decompose $0 \cdot 09328$ milligramme of water, would only decompose $0 \cdot 0627$ milligramme ; this is contrary to Faraday's law.

Thus to explain the decomposition of water by means of a current of an electromotive force of one volt, there would only remain the supposition that the oxygen and hydrogen obtained possess other properties of combination than those which these gases generally have, or, in other terms, would imply that 8 grammes of oxygen, when burning one gramme of hydrogen, produce 22 calories, instead of $34 \cdot 5$, which is absolutely incorrect.

We can therefore affirm that it is impossible to electrolyse water with an electromotive force inferior to $1 \cdot 495$ volt. An electromotive force superior to $1 \cdot 495$ volt will even be required for overcoming the resistances of the conductors and of the bath, which resistances can never be completely done away with.

Quite recently M. Tommasi professed to be able to decompose water with a single cell of zinc-carbon or zinc-copper, with dilute sulphuric acid ; but in order to obtain this result, he made use of suitable electrodes. For instance, when using copper or zinc as anode and platinum as cathode, he very rapidly decomposed water with a single zinc-carbon cell; and when using a platinum anode and a copper cathode he equally obtained a rapid decomposition with a single zinc-copper cell. In these decompositions hydrogen only would be given off, oxygen becoming combined with the positive electrode.

After a long series of experiments, M. Tommasi concluded as follows: 1. One single zinc-copper or zinc-carbon cell, immersed in dilute sulphuric **acid**, does **not decompose water**, according to the theory, if the **two** electrodes **are of platinum** ; 2. In order **to** effect the decomposition of **water, the** positive electrode must **be** constituted **of such a substance that** when submitted **to the influence of an** electric **current it** should **be** capable of combining itself with the oxygen **of the water.**

The explanation of the results obtained **by M.** Tommasi **is** exceedingly simple. **M.** Tommasi assumed **that he was** electrolysing water by means of **a single** cell, whereas he was really using two cells joined in series, and placed one externally to and the other **in** the **interior itself of the voltameter.** .

The zinc-copper cell **was** giving off 19 calories, when at **the same** time **the** copper **at** the positive electrode was getting oxydised, **being transformed** into sulphate, and giving off 28·2 calories. Instead of requiring 34·**5 calories** the dissolution would then only require $31 \cdot 5 - 28 \cdot 2 = 6 \cdot 3$ **calories, a number** greatly inferior **to the 19** calories supplied **by** the zinc-copper cell.

CHAPTER III.

WORK ABSORBED IN ELECTROLYSIS.

General Formulæ—Amount of Work required for the Decomposition of
Water—Electrolysis of a Binary Compound with a Metallic Basis—
Electrolysis of Ternary Compounds—Of the Chemical Actions in an
Electrolyte—Sprague's Law—Limits of Electrolysis—Of the Use of
Soluble Anodes—Practical Determination of the Electromotive Force and
of Metallic Resistances.

GENERAL FORMULÆ.—The work absorbed in an electrolytic
operation is composed of a work corresponding to the heat
developed by the inverted formation of the constituents of the
bath and of the work corresponding to the heat which is absorbed
by the conductor, in the whole length of the circuit.

One of these two expenditures of work is invariable for every
electrolyte and cannot be modified by the practitioner; it is that
which is required by the counter electromotive force; the other
expenditure, which is dependent on the resistance of the circuit,
can, on the contrary, be reduced by means of various devices
such as the use of electrodes of large dimensions, baths of a
greater conductivity, reduction of the distance between the baths
and the generator, use of large size conducting cables, reduction
of the distance between the electrodes, increase of the temperature, &c.

The calculation of the work required for the decomposition
proper is therefore the only one of the two which can be
determined with mathematical correctness; that relating to
the resistances can only be approximately valued for each particular application.

Calling

C the intensity of the current, in amperes;

E the counter electromotive force in volts;

R the total resistance of the circuit in ohms;

w the chemical **work**, in kilogrammetres;
w' the calorific work, in kilogrammetres;
W the total **work in** kilogrammetres.

The expression **of the** work **of decomposition** will be represented by the formula:

$$w = \frac{CE}{g} \text{ kilogrammetres per second;}$$

that of the work absorbed by **the** resistance by:

$$w' = \frac{C^2 R}{g} \text{ kilogrammetres per second;}$$

and that of **the total work by:**

$$W = \frac{CE + RC^2}{g}.$$

If it is proposed to obtain a given weight of decomposed substances, **per hour, the number of amperes** will be immediately determined by the electro-chemical **equivalent** of the electrolyte.

One ampere **liberates** 0·00001036 gramme **of** hydrogen per second, which corresponds to 0·037296 gramme per hour.

In order to estimate the quantity **of** metal liberated **by the** electric current in **one** hour's time, **it is** sufficient to multiply the chemical equivalent of **the** said metal by 0·037296. Reciprocally, when **the** weight **of the metal to** be precipitated **per** hour is known, the number **of amperes required is obtained by** dividing the product **of the chemical equivalent of the metal** multiplied by 0·**037296 into the said weight.**

Calling H e the number of calories **given off by one** chemical equivalent of the dissolution, **the number of volts will** be, as already said: $E = \dfrac{H e}{23}$.

Knowing C and E, the work of the electro-chemical decomposition will be immediately determined. The tables (p. 43 and following) **give the** chemical equivalents and the quantities of calories **given off** in the electrolytes **in** general use. **If the** heat due **to the combination of** the compound substance were not given **in the said tables, it** would be necessary to determine it **by** means **of direct** experiments.

AMOUNT OF WORK NECESSARY FOR THE DECOMPOSITION OF WATER.—We will propose to determine the work corresponding to the decomposition of one kilogramme of water in one hour, for example.

We have seen that the said work was represented by the two expressions

$$w = \frac{C\,E}{g} \text{ and } w' = \frac{C^2\,R}{g}.$$

In this instance we have only to deal with $\dfrac{C\,E}{g}$, that is to say, the work absorbed by the decomposition of the water exclusive of the calorific energy expended by the conductors. One ampere current decomposing $0 \cdot 00009328$ gramme, or $0 \cdot 00000009328$ kilogramme of water, to decompose one kilogramme, $\dfrac{1}{0 \cdot 00000009328}$ amperes will be required; and as the electromotive force is $1 \cdot 4856$ volts, the corresponding work will be

$$w = \frac{1 \cdot 4856}{9 \cdot 81 \times 0 \cdot 00000009328} = 1623430 \text{ kilogrammetres.}$$

In order to verify this number, it is sufficient to calculate the work which the hydrogen and the oxygen contained in one kilogramme of water can develop in combining together. One equivalent of water or 9 grammes corresponds to $34 \cdot 462$ calories, 1 kilogramme will therefore correspond to $\dfrac{34462}{9} = 3829 \cdot 111$ calories, or to $3829 \cdot 11 \times 424 = 1633542$ kilogrammetres (the slight discrepancy existing between these two results arising solely from the neglect, in both the calculations, of some decimals).[*]

The number 1623500 kilogrammetres can be adopted as sufficiently accurate in all industrial calculations. The work being effected in one hour's time, it will correspond to $\frac{1623500}{3600} = 451$ kilogrammetres per second; that is to say, approximately to 6 horse-power. Therefore, one horse-power corresponds to 167 grammes of water decomposed per hour. In practice, the work necessary for the decomposition of water is greatly increased by resistances of all kinds. M. Gramme, who happened to incidentally treat the question of electrolysing water, made some experiments with a certain amount of precision, and with a voltameter of his own construction, and he succeeded in decomposing 60 grammes of water per horse-power per hour. If the calorific work absorbed by the circuit is taken into account, the yield certainly exceeds 50 per cent., since it is equal to $\frac{60}{167}$, or 35 per cent. of the work required for the electrolysis proper. We do not pretend that better results cannot be obtained, but we are not aware of anything more economical having been realised in such a direction, and would give manufacturers thinking of employing electrolysis for the production of pure hydrogen or oxygen the advice of not reckoning upon a greater yield.

So, practically, one horse-power, acting during a period of one hour, decomposes 60 grammes of water, and liberates

6·67 grammes or 74·4 litres of hydrogen,
53·33 grammes or 37·2 litres of oxygen.

ELECTROLYSIS OF A BINARY COMPOUND WITH A METALLIC BASIS.—The preceding calculations in reference to the estima-

---

of a book. Since the electromotive force of polarisation is precisely calculated in respect of the work of combination, there should be no other discrepancy between the two numbers of kilogrammetres than that arising from the neglect of some decimal fractions.

Why take the figure 1·75 as representing the electromotive force of polarisation, accepting M. E. E. Blavier's authority (M. E. E. Blavier has however given a different figure) when direct calculation was so easy? M. Japing questions as to the causes which alter the pure theory; if our book comes under his notice he will understand that pure theory is not altered in the said calculation.

tion of the work required for the decomposition of water, apply
equally to all the electrolytes composed of two constituents.
Taking as an example of a metallic salt, a bath of chloride of
zinc (ZnCl), we will purpose liberating the zinc.

If we refer to the tables given, pp. 49–50, we find :—

Chemical equivalent of zinc	.. ..	32·7	
„ „ chlorine	.. ..	35·5	
„ „ chloride of zinc	.. ..	68·2	
Heat developed by the combination of 1 equiva- lent of chloride of zinc in a state of solution ..	}	56·4 calories.	

With these figures, the calculation is of the greatest simplicity; if 56·4 calories are required to be given off by the
electric current in order to precipitate 32·7 grammes of zinc;
to precipitate 1 kilogramme (1000 grammes) the following
quantity would be required, or $\dfrac{56\cdot4 \times 1000}{32\cdot7} = 1721$ calories ;
thus $w = 1721 \times 424 = 729704$ kilogrammetres. This corre-
sponds, for 1 kilogramme per hour, to

$$w = \frac{729704}{3600 \times 75} = 2\cdot7 \text{ horse-power.}$$

The number of volts absorbed in that work is, as we have
already said, $E = \dfrac{H\,e}{23} = 2\cdot45$ volts. The electro-chemical
equivalent of zinc being 0·000332 gramme, the intensity C, in
amperes, of a current capable of liberating 1 kilogramme of
zinc per hour will be :

$$C = \frac{1000}{0\cdot0003332 \times 3600} = 833 \text{ amperes.}$$

The work w as regards the amperes and volts, will, of course,
be the same as regards the equivalents and the calories of
combination. Thus :

$$w = \frac{833 \times 2\cdot45}{9\cdot81} = 208\cdot1 \text{ kilogrammetres, or } 2\cdot7 \text{ horse-power.}$$

If we wished to correctly calculate the number of volts of
the electric source, and consequently the work required for the
precipitation of one kilogramme of zinc per hour, we should

have to successively determine the calorific work absorbed by
the resistance of the complete **circuit,** the mechanical work
absorbed by friction **in** the different. parts of the machine, the
work necessary for the transport of the ions in the electrolyte,
the loss of work due to the polarisation of the electrodes, and
finally, the expenditure of **energy due to all the** secondary
actions taking place **in** the **bath. It** is impossible to integrally effect all these calculations; the work absorbed **by the**
resistance of the circuit can only **be** approximately **estimated,**
and **a** coefficient **of the** mechanical efficiency of the **motive**
power admitted. **As to the other** causes of losses **of motive**
power, they **can be estimated in** block **in** each particular
case, according **to the** magnitude of the electromotive forces
used, and to the nature of the secondary actions which **are liable**
to take place; but these estimations are only endowed with real
interest on **the** condition **of** their **having been** preceded by
preliminary experiments.

In order to determine **the calorific work of the** circuit, it is
necessary **to know :**

(1) The specific **resistance of the electrolyte ;**
(2) The distance between the electrodes;
(3) The surface of the anodes and the cathodes ;
(4) The resistance of the conductors ;
(5) **The** internal **resistance of the** dynamo.

In **the** case of **a solution of** chloride of zinc, we can admit
a specific resistance of **the liquid equal** to $2 \cdot 14$ **ohms, which**
corresponds to $\dfrac{2 \cdot 14}{10000} = 0 \cdot 000214$ ohm **per square metre (the**
specific resistance **being that of** one **cubic centimetre between**
two parallel faces) for **a** thickness **of one centimetre of** the
liquid. We can **admit** that **the** electrodes **are at a** distance of
$0 \cdot 1$ metre, and that the immersed surfaces of the anodes and the
cathodes **are** respectively **of** one square metre for 200 amperes.
The intensity being 833 amperes, the total surface will be
4·16 square metres. The resistance R" **of the** liquid can then
be calculated **by means of** Ohm's formula :

$$R'' = \frac{0 \cdot 000214 \times 10}{4 \cdot 16} = 0 \cdot 000512 \text{ ohm.}$$

D

The resistance of the metallic conductors can bo estimated at $0 \cdot 0014$ ohm; it is that of a bar of copper 10 metres long and 1 square centimetre in section. To illustrate a case, we will suppose that the internal resistance of the dynamo is equal to the tenth part of the total external resistance, and that its mechanical efficiency is 90 per 100, a result which is often realised in practice.

The resistance R' created by the counter electromotive force is

$$R' = \frac{E}{C} = \frac{2 \cdot 45}{833} = 0 \cdot 003 \text{ ohm.}$$

The internal resistance of the dynamo, as well as that due to friction, may be estimated at $0 \cdot 0006$ ohm. Then it can be admitted that a resistance of $0 \cdot 002$ ohm is due to the polarisation of the electrodes, and to other causes of a varied nature which necessitate an expenditure of energy, and which have not been calculated; the total resistance of the circuit is thus shown:

$$R = 0 \cdot 003 + 0 \cdot 000512 + 0 \cdot 0014 + 0 \cdot 0006 + 0 \cdot 002$$
$$= 0 \cdot 007512 \text{ ohm.}$$

The number of amperes 833 remaining constant, the electromotive force required to overcome this total resistance will be:

$$E = \frac{C}{R} = 833 \times 0 \cdot 007512 = 6 \cdot 25 \text{ volts.}$$

The proportion between the work actually required for liberating the zinc and that absorbed in the electrolysis of the chloride of zinc is therefore

$$\frac{2 \cdot 45}{6 \cdot 25} = 0 \cdot 40.$$

The primitive work being $2 \cdot 7$ horse-power, the total work is in reality $\frac{2 \cdot 7}{0 \cdot 40} = 6 \cdot 75$ horse-power.

In order to extract one kilogramme of zinc per hour from a bath of chloride of zinc, it will thus be necessary to have:

(1) An engine of about 7 horse-power;

(2) A dynamo electric machine giving a current of 833

amperes, with a total electromotive force of 6·25 volts (about 5·6 volts at the terminals).

If the conductor had a resistance ten times smaller than the one which we have admitted, and if the dissociation of the bath was not accompanied by any chemical action, the expenditure of the motive power could be reduced by about 20 per cent., or, what amounts to the same, the return could be increased from 0·40 to 0·60. It is therefore important to reduce to a minimum the resistance of the conductors, and to avoid, as much as possible, the polarisation of the electrodes.

As we have already mentioned, the number of amperes is invariable, whatever the passive metallic resistances or the secondary actions of the current may be. This is an absolute consequence of Faraday's law.

With a given dynamo electric machine, the electromotive force may be increased or reduced by varying the speed of the induced bobbin; but it is practically impossible to exceed a certain number of amperes without running the risk of burning the insulating substances with which both the induced and inducting wires are covered.[*]

ELECTROLYSIS OF TERNARY COMPOUNDS.—In order to explain the electrolysis of a ternary compound, we will as an illustration take the sulphate of copper for example, and follow step by

---

[*] It would even be well, when working out the details of a machine, not to exceed a certain number of amperes, otherwise purely fancy results are obtained, which are of no practical value. It is exactly as if it were proposed to try, with steam at a pressure of say 500 lbs., a steam engine constructed to work with a pressure of 50 lbs. to the square inch; the stuffing-boxes and pistons would let the steam escape, the connecting-rods would bend or break, &c., &c.

In 1872 M. Gramme had combined, for Messrs. Christofle, a machine capable of depositing 600 grammes of silver per hour, and the wire of which was of a sufficiently large section for a current of 200 amperes, whereas the current required for precipitating the metal was only 150 amperes. This machine gave excellent results. It was later on forwarded to Birmingham, where it deposited 1200 grammes of silver per hour. This led the late M. Werdermann, who conducted the experiments, to the conclusion that English manufacturers made a better use of the electric current than French ones. Nothing was less correct; the English baths had merely less resistance owing to the anodes and cathodes being placed in closer proximity, which is far from being a good plan for obtaining a good deposition; and the machine was run so as to give a current of 300 amperes, with the result that it heated very much, and could not be kept running for a long time without stoppages.

step M. Berthelot in the calculation of the useful and secondary effects.

When a bath of sulphate of copper is traversed by an electric current, one equivalent of copper is deposited on the negative pole, whereas the other constituents, such as the oxygen which is liberated and the dilute sulphuric acid which gradually accumulates, are transported to the positive pole. Instead of liberating the elementary constituents, as was the case with chloride of zinc, the reaction liberates the metal on the one part, and the system SO_4 on the other part:

$$SO_4 Cu = \overset{+}{Cu} + O + \overline{SO_3}.$$

SO_4 is gradually resolved into oxygen which is liberated, and into sulphuric acid, which remains dissolved in the excess of water.

Such is the normal type of all decompositions of a ternary, or even of a more complicated salt.

The chemical work performed in the course of this decomposition is measured by the sum of the quantities of heat developed firstly when the copper combines with the oxygen and then when the oxide of copper and the dilute sulphuric acid combine together, or :

$$19 + 9 \cdot 2 = 28 \cdot 2 \text{ calories.}$$

M. Berthelot remarks, with reason, that an electric current produces besides, the heating of the liquors to a certain extent, and causes a certain displacement of substances comprising not only sulphuric acid, oxygen, and copper, but also a portion of the mass of the sulphate of copper; the relative proportion and the laws of such accessory work are not well known.

The electric action is such as we have described it, at the beginning of the operation; but as soon as the latter has begun, the liquor contains a certain amount of dilute sulphuric acid, and particularly in the vicinity of the positive pole. From that moment the acid becomes electrolysed at the same time that the sulphate of copper, that is to say hydrogen is liberated at the negative pole. One portion of that hydrogen is freed;

another portion reduces a certain quantity of sulphate of copper, with a precipitation of metallic copper.[*]

The more constituents there are in the composition of compound substances, the more causes of secondary actions there are in a bath, and the comparatively smaller is the yield for any given one of the said constituents. Thus with chloride of zinc we adopt a yield of 40 per 100, whereas with sulphate we will not admit a yield of more than 30 per 100.

Now as to the calculation of the electrolytical work.

One equivalent of sulphate of copper is composed of 32 grammes of oxygen, 16 grammes of sulphur, and $31 \cdot 8$ grammes of copper; $28 \cdot 2$ calories must therefore be expended for the precipitation of $31 \cdot 8$ grammes of copper. One kilogramme would require

$$\frac{1000 \times 28 \cdot 2}{31 \cdot 8} = 886 \text{ calories}$$

and $W = 886 \times 424 = 375664$ kilogrammetres.

To effect this work in one hour's time, with a yield of 30 per 100, an engine power of

$$\frac{375664}{0 \cdot 30 \times 3600 \times 75} = 4 \cdot 33 \text{ horse-power will be required.}$$

Theoretically $1 \cdot 30$ H.P. is sufficient. The number of amperes must be

$$\frac{1000}{0 \cdot 00001036 \times 31 \cdot 8 \times 3600} = 840.$$

The number of volts is

$$\frac{4 \cdot 33 \times 75 \times 9 \cdot 81}{840} = 3 \cdot 86.$$

So that $1 \cdot 16$ volt would have been sufficient for overcoming the work of decomposition proper.

OF THE CHEMICAL ACTIONS IN AN ELECTROLYTE.—If it is always easy to calculate the number of amperes required for the liberation of a given weight of metal in a given electrolyte, and when the chemical actions are known as also the useful work for overcoming the counter electromotive force, it is, on the con-

* Berthelot, 'Essai de mécanique chimique,' t. ii. p. 330.

trary, often a very difficult thing to estimate all the chemical phenomena which *accompany* a decomposition; and notwithstanding the tables setting forth the chemical and calorific equivalents, it is only with great difficulty that the accurate total of all the energies contributing to the electrolytic phenomena can be ascertained. It is therefore desirable, for every application, not to undertake an industrial operation before having previously conducted some laboratory experiments and carefully examined the actions which take place and the definite results to which they give rise.

· To complete this chapter and assist the operator in his researches, we will give a few summary directions respecting the use of soluble anodes, the law of the lesser energy of Sprague, the limits of electrolysis of Berthelot, and some details on M. Marchese's calculations for estimating the resistances of an electric circuit.

SPRAGUE'S LAW.—M. Sprague's law on electrolysis can be formulated as follows:

The substances which in becoming freed absorb the smallest quantity of intrinsic energy are those which are liberated at the electrodes.

This amounts to saying that any molecule, electrolyte or not, in the neighbourhood of the points at which the current enters the electrolyte or goes out of it, will be ranged within the polar circuit, provided a portion of the said molecule can combine itself with the veritable ion which is turned towards the electrode and the other portion may constitute a free molecule absorbing less energy than the ion of the electrolyte would itself absorb from becoming liberated, or else provided the veritable ion could be introduced into the other molecule with a less expenditure of energy than that required to allow of its constituting a free molecule. The substances liberated can thus be recombined anew by those which are in contact with the electrodes, as they are no more simply separated in a state of ions or radicals resulting from molecules previously existing, although this latter action is the fundamental type and general action of electrolysis.

This new conception establishes an analogy between the effects of electricity passing through a liquid and those due to the process of heating during the dissociation. In these two cases, effectively, the molecules become grouped according to the various forces in action : degrees of heat in these ; electromotive forces in those.

Sprague's law is the converse of that which M. Berthelot designated by the name of " principle of the maximum work," and which he enumerated as follows : " Any chemical change obtained without the interference of a foreign energy has a tendency towards the production of the substance or the system of substances which develop the greatest heat."

LIMITS OF ELECTROLYSIS.—M. Berthelot has carried on some interesting experiments concerning the decomposition of a certain number of electrolytes, and he has deduced from them the following conclusions :

1. In the generality of cases, the decomposition of electrolytes takes place as soon as the smallest sum of the necessary energies—that is to say, in accordance with the quantities of heat —is present. The said sum is calculated taking into account all the reactions effected during the passage of the current, without it being necessary to make a distinction between the reactions called primitives and the reputed secondary reactions.

2. The limit of forces is uncertain whenever polarisation arises. In order to take this conclusion into account and ascertain if it is in accordance with the common law, it would be necessary to know the real nature of the compound substances which it brings into action, and the heat due to their combination.

We have condensed in the table given, page 58, the experiments of M. Berthelot. They have the great merit of throwing light on a few points concerning the real actions taking place in electrolytic baths, which had until then remained obscure.

OF THE USE OF SOLUBLE ANODES.—In nearly all the industrial applications of electrolysis, the platinum or carbon anodes are replaced by soluble anodes ; the result is to maintain the chemical neutrality of the liquors and to suppress the work necessitated by the decomposition.

In the case of sulphate of copper, if the anode is a plate of copper, it will receive some oxygen and sulphuric acid which will attack it and form some sulphate of copper which will be dissolved in the bath. If, therefore, the cathode increases in weight, the anode will decrease in the same proportion. The work of decomposition will be exactly compensated by that of recomposition, and the electrical energy will be expended only by the conductor and by some small secondary actions.

Notwithstanding this arrangement, the liquor will, after a certain time, be subject to some changes in its degree of saturation; the sulphate will gradually decrease round the negative, and concentrate round the positive pole. The remedy for this want of homogeneity in the body of the liquor consists in producing a continuous mechanical movement in the vat, so as to constantly mix all the parts which are in a state of dissolution.

When treating of the refining of metals we shall have an opportunity of examining the work expended when soluble anodes are used, and we shall see that this work is far from being nil. The necessity of employing baths of very great resistance and to keep the electrodes at a certain distance from each other in order to obtain the metal in a state of absolute purity create considerable resistances which often absorb more than 2 volts.

This comparatively large electromotive force has the inconvenience of producing decomposition of the water, which would not take place under $1\frac{1}{2}$ volt, and of constituting a very important waste of work. The phenomena of polarisation and of inversion of currents, which are so prejudicial to industrial electrolytic operations, are also caused by the decomposition of the water.

PRACTICAL DETERMINATION OF THE COUNTER ELECTROMOTIVE FORCE AND OF METALLIC RESISTANCES. — M. E. Marchese, an engineer of Genoa, who makes his speciality of the treatment of sulphide of copper by electricity, has quite recently published a note on the determination of the resistances of the circuit. We will consider its principal points.

The difference of potential observed between the two electrodes of an electro-chemical bath arises out of the counter electromotive force which is necessitated by the dissociation of the electrolyte and of the electromotive force absorbed by the metallic resistances of every kind. In order to obtain an

economical efficiency, it is useful to know the proportion existing between these two electromotive forces, or, what amounts
to the same, between their two corresponding resistances. The
determination, by means of calculation, of these two quantities,
is not an easy matter, as the chemical actions are often of a very
complex nature, and the elements of the metallic resistance are
extremely varied and subject to great alterations even during
the course of a single operation.

If the intensity C is measured in amperes, the fall of potential E between the electrodes, in volts, the resistance evidently
is $R = \dfrac{E}{C}$.

Calling E' the counter electromotive force, and R' the
metallic resistance of the bath, we shall have $R' = \dfrac{E - E'}{C}$.
In the industrial practice and for every electrolyte, E' can be
considered as being constant, and R' as variable, owing to the
differences of heat, saturation, acidity, &c., which, although they
may be slow in manifesting themselves, nevertheless occur
with certainty for the whole duration of the operation. It can,
however, be assumed that R' is constant during a short interval
of time.

M. Marchese advises estimating **E** and C on the electrolyte
which is under observation, then to alter the regimen by
increasing or reducing the speed of the dynamo machine and
make some new observations.

Thus at a normal speed, we have, for example:

$$\frac{E - E'}{C} = R.$$

For a greater speed we obtain

$$\frac{E + e - E'}{C + c} = R'.$$

We deduce from these two equations:

$$E' = E - \frac{e\,C}{c}$$

$$R' = \frac{e}{c}.$$

If, for example, $E = 0 \cdot 9$ volt and $C = 150$ amperes. If we suppose that after the acceleration of speed we find :

$E + e = 1 \cdot 1$ volt, from which $e = 0 \cdot 2$ volt.

$C + c = 250$ amperes, from which $c = 100$ amperes.

We shall have for the counter electromotive force :

$$E' = 0 \cdot 9 - 0 \cdot 2 \frac{150}{100} = 0 \cdot 6 \text{ volt,}$$

and for the metallic resistance of the bath :

$$R' = \frac{0 \cdot 2}{100} = 0 \cdot 002 \text{ ohm.}$$

The resistance of the conductors and of the machine itself can be correctly ascertained ; it is therefore very easy to calculate the total resistance and the power required for the treatment of the electrolytic substances which it is proposed to separate.

In order to obtain as high an economical result as possible, it will be useful to determine from time to time both the counter electromotive force which is developed between the electrodes and the metallic resistance of the bath.

CHAPTER IV.

TABULAR DATA.

Resistance of Metals and usual Alloys—Relative Conductivity of Metals—Influence of the Temperature on the Resistance of Metals and Pure Copper—Resistance of Pure Copper Wires and Bars—Conductivity of various Coppers—Resistance of various Saline Solutions—Resistance of various Acids—Chemical and Electro-chemical Equivalents—Heat Developed by the Formation of the principal Binary and Ternary Compounds—Motive Power required for the separation of various Metals from their Solutions—Limits of Electrolysis, Experiments of M. Berthelot.

RESISTANCE OF METALS AND USUAL ALLOYS AT 0° CENTIGRADE.

(Matthiessen.)

Name of Metals.	Resistance of a cubic centimetre between its opposite faces.	Percentage of increase in resistance per degree centigrade.	Observations.
	Microhms.		These figures,
Silver, annealed	1·521	0·377	which have been
„ hard-drawn	1·652	..	published in the
Copper, annealed	1·616	0·388	'Transactions of the
„ hard-drawn	1·652	..	British Association,'
Gold, annealed	2·081	0·365	have been, until now,
„ hard-drawn	2·118	..	accepted without dis-
Aluminium, annealed	2·945	..	cussion as the prac-
Zinc, pressed	5·689	..	tical measures.
Platinum, annealed	9·158	..	
Iron, annealed..	9·825	0·630	
Nickel, annealed	12·600	..	1 microhm =
Tin, pressed	13·360	0·365	0·000001 ohm.
Lead, pressed	19·850	0·387	
Antimony, pressed	35·900	0·389	
Bismuth, pressed	132·700	0·354	
Mercury, liquid	99·740	0·072	
1 Silver, 2 Platinum	24·660	0·031	
German silver	21·170	0·044	
2 Gold, 1 Silver	10·990	0·065	

RELATIVE CONDUCTIVITY OF METALS. COMPARED TO SILVER AND PURE COPPER. (Lazare Weiller.)

Name of Metals.	Conductivity.	Observations.
1. Silver, pure..	100	These experiments have
2. Copper, pure	100	been conducted with a series
3. Copper, pure, super refined and crystallised	99·9	of bars especially prepared for the purpose. These said
4. Silicium bronze (telegraphic)	98	bars have been molten at a
5. Copper and silver alloy at 50 per cent.	86·65	uniform diameter of about
6. Gold, pure	78	13 millimetres. They have
7. Silicic copper (with 4 per 100 of silicon)	75	been cut so as to show the
8. Silicic copper (with 12 per 100 of silicon)	54·7	grain of the metal, and the detached portions have then
9. Aluminium, pure	54·2	been drawn into wires.
10. Tin, containing 12 per 100 of sodium	46·9	It is on the wires so obtained that the said experi-
11. Silicium bronze (telephonic)	35	ments have been carried out
12. Plombiferous copper, with 10 per 100 of lead	30	and of which the results are
13. Zinc, pure	29·9	given in the table.
14. Phosphor bronze (telephonic)	29	As regards those alloys
15. Silicium brass, with 25 per 100 of zinc	26·49	which can neither easily be
16. Brass, with 35 per 100 of zinc.. ..	21·15	drawn or rolled, such as cer-
17. Phosphide of tin..	17·7	tain phosphides or silicides,
18. Gold and silver alloy, 50 per cent. ..	16·12	the measurements have been
19. Swedish iron	16	taken direct from the bars
20. Pure tin of Banca	15·45	according to the method of
21. Antimonous copper	12·7	Sir W. Thomson.
22. Aluminium bronze, 10 per 100 ..	12·6	The measurements have
23. Siemens steel	12	been taken by means of a
24. Platinum, pure	10·6	Wheatstone bridge with a
25. Amalgam of cadmium, with 15 per 100 of cadmium	10·2	sliding index, a differential galvanometer, and a battery
26. Mercurial bronze, Drosnier	10·14	of four cells.
27. Arsenical copper, with 10 per 100 of arsenic	9·1	
28. Lead, pure	8·88	
29. Bronze, with 20 per 100 of tin.. ..	8·4	
30. Nickel, pure	7·89	
31. Phosphor bronze, with 10 per 100 of tin	6·5	
32. Phosphide of copper, with 9 per 100 of phosphorus	4·9	
33. Antimony	3·88	

INFLUENCE OF THE TEMPERATURE ON THE RESISTANCE OF METALS. (Matthiessen.)

According to the formula $R = r (1 + at + b t^2)$,
R resistance at $t°$; r resistance at $0°$; t centigrade degrees.
Value of numerical coefficients a and b for a few metals.

Metals.	Value of a.	Value of b.
Very pure metals	+ 0·0038240	+ 0·000001260
Mercury	+ 0·0007485	− 0·000000398
German silver	+ 0·0004433	+ 0·000000152
Silver and platinum alloy	+ 0·0003100	..
Silver and gold alloy	+ 0·0006999	− 0·000000062

CONDUCTIVITY OF METALS (Matthiessen).
Coefficients for the temperature: t in degrees centigrade.

Metals.	Coefficients.
Silver	$C = 100 - 0.38278\,t + 0.0009848\,t^2$
Copper	$C = 100 - 0.38701\,t + 0.0009009\,t^2$
Gold	$C = 100 - 0.36745\,t + 0.0008443\,t^2$
Zinc	$C = 100 - 0.37047\,t + 0.0008274\,t^2$
Cadmium	$C = 100 - 0.36871\,t + 0.0007575\,t^2$
Tin	$C = 100 - 0.36029\,t + 0.0006136\,t^2$
Lead	$C = 100 - 0.38756\,t + 0.0009146\,t^2$
Arsenic	$C = 100 - 0.38996\,t + 0.0008879\,t^2$
Antimony..	$C = 100 - 0.39826\,t + 0.0010364\,t^2$
Bismuth	$C = 100 - 0.35216\,t + 0.0005728\,t^2$
Average ..	$C = 100 - 0.37047\,t + 0.0008340\,t^2$

INFLUENCE OF THE TEMPERATURE ON THE RESISTANCE AND THE CONDUCTIVITY OF PURE COPPER.

Temperature Centigrade.	Resistance.	Conductivity.	Temperature Centigrade.	Resistance.	Conductivity.	Temperature Centigrade.	Resistance.	Conductivity.
0	1.00000	1.00000						
1	1.00381	0.99624	11	1.04199	0.95970	21	1.08164	0.92452
2	1.00756	0.99250	12	1.04599	0.95603	22	1.08553	0.92121
3	1.01135	0.98878	13	1.04990	0.95247	23	1.08954	0.91782
4	1.01515	0.98508	14	1.05406	0.94893	24	1.09356	0.91445
5	1.01896	0.98139	15	1.05774	0.94541	25	1.09763	0.91110
6	1.02280	0.97771	16	1.06168	0.94190	26	1.10161	0.90776
7	1.02663	0.97406	17	1.06563	0.93841	27	1.10567	0.90443
8	1.03048	0.97042	18	1.06959	0.93494	28	1.11972	0.90113
9	1.03435	0.96679	19	1.07356	0.93148	29	1.11382	0.89784
10	1.03822	0.96319	20	1.07742	0.92814	30	1.11782	0.89457

RESISTANCE OF ANNEALED PURE COPPER WIRES AND BARS AT 0° CENT.

Diameter in Millimetres.	Section in Square Millimetres.	Weight per Metre in Grammes.	Length per Kilogramme in Metres.	Resistances.		
				Ohms per Metre.	Metres per Ohm.	Ohms per Kilogramme.
0.5	0.1963	1.74	576.0	0.081	12.305	46.81
1.0	0.7854	6.99	144.0	0.0203	48.872	2.95
1.5	1.7670	15.75	63.0	0.0091	109.75	0.574
2.0	3.1420	27.95	36.0	0.0051	195.15	0.185
2.5	4.9087	43.70	23.0	0.0032	308.60	0.0745
3.0	7.0690	62.93	16.0	0.0023	439.10	0.0365
3.5	9.6214	85.67	11.7	0.00165	605.00	0.0193
4.0	12.566	111.9	8.94	0.00129	777.00	0.01150
4.5	15.900	141.8	7.05	0.001011	989.00	0.00713
5	19.630	174.0	5.76	0.000810	1231.00	0.00468
6	28.276	251.7	4.00	0.000580	1756.3	0.00232
7	38.486	342.7	2.92	0.000412	2420.0	0.00120
8	50.264	447.3	2.23	0.000321	3008.0	0.000718
9	63.601	567.0	1.76	0.000253	3956.0	0.000455
10	78.540	699.0	1.44	0.000203	4878.2	0.000295

RESISTANCE OF ANNEALED PURE COPPER WIRES AND BARS—*continued.*

Diameter in Milli-metres.	Section in Square Milli-metres.	Weight per Metre in Grammes.	Length per Kilogramme in Metres.	Resistances.		
				Ohms per Metre.	Metres per Ohm.	Ohms per Kilogramme.
11	94·250	840·0	1·20	0·000167	5854·0	0·000202
12	113·10	1007·0	0·993	0·000145	7025·2	0·000145
13	133·45	1190·0	0·847	0·000118	8293·0	0·0001035
14	153·94	1371·0	0·730	0·000103	9680·0	0·0000750
15	176·70	1575·0	0·630	0·0000910	10975·0	0·0000574
16	201·04	1789·0	0·556	0·0000804	12032·0	0·0000449
17	227·70	2001·0	0·500	0·0000692	14450·0	0·0000346
18	254·41	2268·0	0·440	0·0000632	15824·0	0·0000278
19	283·32	2520·0	0·400	0·0000556	17561·0	0·0000227
20	314·20	2795·0	0·360	0·0000510	19515·0	0·0000185
25	490·87	4370·0	0·230	0·0000324	30860·0	0·00000715
30	706·90	6293·0	0·160	0·0000230	43907·0	0·00000365
35	962·14	8567·0	0·117	0·0000165	60500·0	0·00000193
40	1256·6	11184·0	0·0894	0·0000129	77700·0	0·000001150
45	1590·0	14175·0	0·0705	0·00001011	98900·0	0·000000713
50	1963·0	17400·0	0·0576	0·00000810	123050·0	0·000000468

CONDUCTIVITY OF COPPER ALLOYED WITH FOREIGN SUBSTANCES.
PURE COPPER = 100. (Matthiessen.)

Substances alloyed with Pure Copper.	Conducti-vity com-pared to Copper.	Tempera-ture in Cent.	Substances alloyed with Pure Copper.	Conducti-vity com-pared to Copper.	Tempera-ture in Cent.
		°			°
0·5 per 100 of carbon .	77·87	18·3	0·48 per 100 of iron ..	35·92	11·2
0·18 ,, sulphur	92·08	19·4	1·66 ,, ,, ..	28·01	13·1
0·13 ,, phosphorus	70·34	20·0	1·33 ,, tin ..	50·44	16·8
0·95 ,, ,,	24·16	22·1	2·52 ,, ,, ..	33·93	17·1
2·5 ,, ,,	7·52	17·5	4·9 ,, ,, ..	20·24	14·4
Copper with traces of arsenic	60·08	19·7	1·22 ,, silver..	90·34	20·7
2·8 per 100 of arsenic.	13·66	19·3	2·45 ,, ,, ..	82·52	19·7
5·4 ,, ,,	6·42	16·8	3·5 ,, gold ..	67·94	18·1
Copper with traces of zinc	88·41	19·0	10·0 ,, aluminium	12·68	14·0
1·6 per 100 of zinc ..	79·37	16·8	0·31 ,, antimony	64·5	12·0
3·2 ,, ,, ..	59·23	10·3	0·29 ,, lead ..		

RELATIVE CONDUCTIVITY OF COPPER OF VARIOUS EXTRACTIONS. (Matthiessen.)

Substances.	Conduc-tivity.	Tempera-ture.
1. Spanish, from Rio-Tinto, with 2 per 100 of arsenic, traces of lead, nickel, suboxide	14·24	14·8 °
2. Russian, from Demidoff, with traces of arsenic, iron, nickel, suboxide	59·34	12·7
3. Tough cake, extraction not mentioned, with traces of lead, iron, nickel, antimony, suboxide	71·03	17·3
4. Best selected, extraction not mentioned, with traces of iron, nickel, antimony, suboxide	81·35	14·2
5. Australian, from Burra-Burra, traces of iron and suboxide ..	88·86	14·0
6. American, from Lake Superior, traces of iron, suboxide, and 0·03 per 100 of silver	92·57	15·0

INFLUENCE OF THE **SUBOXIDE.** (Matthiessen.)

Wire, hard drawn.	Process of Preparation.	Conductivity.	Temperature.
Copper, pure	Oxide reduced by the hydrogen	93·00	18·6
"	Galvanic copper not melted	93·46	20·2
"	Copper of commerce, not melted	93·02	18·4
"	The same after melting in hydrogen	92·76	19·3
"	The same with a current of hydrogen through the molten metal .. ..}	92·99	**17·5**
	REMARK.—The conductivity increases by about 2·5 per 100 when the wires are annealed.		
	Copper containing some suboxide, the proportion of which has not been correctly determined ..}	73·32	**19·5**
	Galvanic copper, extracted from a dense ingot, melted under the coal and run in gas ..}	93·3	**12·8**
	Copper from a porous ingot of the same metal as before, but run into a mould under ordinary conditions .. ..}	94·8	**13**
	Galvanic copper cemented with coal, and containing silicon with traces of phosphorus and iron .. ..}	62·8	13

RESISTANCE OF CERTAIN LIQUIDS.

Sodic Chloride (Sea Salt).		Potassic Nitrate (Saltpetre).	
Weight contained in 100 Grammes of Water.	Resistance in Ohms at 18° Centigrade.	Weight contained in 100 Grammes of Water.	Resistance in Ohms at 18° Centigrade.
25·8758	0·38106	18·9167	0·53018
24·4033	0·36917	13·7647	0·70436
20·9787	0·40647	10·4840	0·86019
17·0174	0·45275	6·6079	1·24140
10·4525	0·66172	4·3964	2·11790
6·6957	0·99070	1·5452	4·06427
3·6880	1·56940		
1·7177	3·54370		

SPECIFIC RESISTANCE OF SOLUTIONS OF SULPHATE OF COPPER AT 10° CENTIGRADE. (Ewing and MacGregor.)

Density.	Specific Resistance.	Density.	Specific Resistance.
1·0167	164·4	1·1386	**35·0**
1·0216	134·8	1·1432	**34·1**
1·0318	98·7	1·1679	**31·7**
1·0622	59·0	1·1823	**30·6**
1·0858	47·3	1·2051 (saturated)	**29·3**
1·1174	38·1		

SPECIFIC RESISTANCE OF SOLUTIONS OF SULPHATE OF COPPER. (Fleeming Jenkin.)

Sulphate of Copper.	Temperature Centigrade.	14°	16°	18°	20°	24°	28°	**30°**
8 parts	Water : 100 parts	45·7	43·7	41·9	40·2	37·1	34·2	32·9
12 "	100 "	36·3	34·9	33·5	32·2	29·9	27·9	27·0
16 "	100 "	31·2	30·0	28·9	27·9	26·1	24·6	24·0
20 "	100 "	28·5	27·5	26·5	25·6	24·1	22·7	22·2
24 "	100 "	26·9	25·9	24·8	23·9	22·2	20·7	20·0
28 "	100 "	24·7	23·4	22·1	21·0	18·8	16·9	16·0

WIEDEMANN'S EXPERIMENTS (1856).

Solution of Sulphate of Copper: SO_3, $CuO + 5$ HO at $14°$ C.

			grammes.					ohms.
31·17	grammes, or		19·95	anhydrous sulphate in one litre	..	..		124·3
62·34	,,		39·90	,,	,,	..	..	70·2
77·92	,,		49·87	,,	,,	..	..	59·9
93·51	,,		59·85	,,	,,	..	..	54·8
124·68	,,		79·80	,,	,,	..	..	45·7
155·85	,,		99·75	,,	,,	..	..	39·9
187·02	,,		119·70	,,	,,	..	..	36·3

SPECIFIC RESISTANCE OF SULPHURIC ACID SOLUTIONS. (Fleeming Jenkin.)

Densities.	Temperatures.							
	$0°$	$4°$	$8°$	$12°$	$16°$	$20°$	$24°$	$28°$
1·10	1·37	1·17	1·04	0·925	0·845	0·786	0·737	0·709
1·20	1·33	1·11	0·926	0·792	0·666	0·567	0·486	0·411
1·25	1·31	1·09	0·896	0·743	0·624	0·509	0·434	0·358
1·30	1·36	1·13	0·94	0·79	0·662	0·561	0·472	0·394
1·40	1·69	1·47	1·30	1·16	1·05	0·964	0·890	0·839
1·50	2·74	2·41	2·13	1·89	1·72	1·61	1·32	1·43
1·60	4·32	4·16	3·62	3·11	2·75	2·46	2·21	2·02
1·70	9·41	7·67	6·25	5·12	4·23	3·57	3·07	2·71

SPECIFIC RESISTANCE OF SULPHURIC ACID SOLUTIONS. (Matthiessen.)

Specific Gravity.	Percentage of Sulphuric Acid in Weight.	Temperature Centigrade.	Resistance.
1·003	0·5	16·1°	16·010
1·018	2·2	15·2	5·470
1·053	7·9	13·7	1·884
1·080	12·0	12·8	1·368
1·147	20·8	13·6	0·960
1·190	26·4	13·0	0·871
1·215	29·6	12·3	0·830
1·225	30·9	13·6	0·862
1·252	34·3	13·5	0·874
1·277	37·3	..	0·930
1·348	45·4	17·9	0·973
1·393	50·5	14·5	1·086
1·493	60·6	13·8	1·549
1·638	73·7	14·3	2·786
1·726	81·2	16·3	4·337
1·827	92·7	14·3	5·320

SPECIFIC RESISTANCE OF NITRIC ACID (Density = 1·36).

Temperature.	$2°$	$4°$	$8°$	$12°$	$16°$	$20°$	$24°$	$28°$
	1·94	1·83	1·65	1·50	1·39	1·30	1·22	1·18

CONDUCTIVITY OF SOLUTIONS OF SEA SALT AT 13° C. (Matthiessen.)

Copper, pure = 100,000,000.	Sea Salt concentrated.	With an equal Volume of Water.	With 2 Volumes of Water.	With 3 Volumes of Water.
Conductivities ..	31·52	23·08	17·48	13·58

SPECIFIC RESISTANCE OF VARIOUS LIQUIDS.

Liquids.	Resistance.
	ohms.
Potassic chloride (27·6 grammes in 500 gr. of water)	1·13
„ „ double volume of water	2·16
„ „ quadruple „	3·92
Sodic chloride (27·6 grammes in 500 gr. of water)	1·13
„ „ double volume of water	2·91
Calcic chloride, dissolved, density 1·04	1·30
Magnesic chloride	1·30
Zinc chloride	2·14

CHEMICAL AND ELECTRO-CHEMICAL EQUIVALENTS.

Name of Substances.	Symbols.	Equivalents.		Weights decomposed by one ampere in one hour.
		Chemical.	Electro-Chemical.	
			mgr.	grammes.
Hydrogen	H	1	0·01036	0·0375
Aluminium	Al	13·7	0·1425	0·5137
Antimony	Sb	122	1·2688	4·575
Arsenic	As	75	0·7800	2·8125
Barium	Ba	68·5	0·7124	2·5687
Bismuth	Bi	210	2·184	7·875
Boron	B	11	0·1144	0·4125
Bromine	Br	80	0·832	3
Cadmium	Cd	56	0·5824	2·095
Calcium	Ca	20	0·208	0·75
Carbon	C	6	0·0624	0·2250
Chlorine	Cl	35·5	0·3692	1·3312
Chromium	Cr	26·2	0·2725	0·9825
Cobalt	Co	29·5	0·3068	1·1062
Copper	Cu	31·8	0·3307	1·1925
Fluorine	F	19	0·1976	0·7125
Gold	Au	98·3	1·0223	3·6862
Iodine	I	127	1·3208	4·7625
Iron	Fe	28	0·2912	1·05
Lead	Pb	103·5	1·0764	3·8812
Magnesium	Mg	12·2	0·1269	0·4575
Manganese	Mn	27·5	0·286	1·0312
Mercury	Hg	100	1·036	3·75
Nickel	Ni	29·5	0·3068	1·1062

CHEMICAL AND ELECTRO-CHEMICAL EQUIVALENTS—*continued.*

Name of Substances.	Symbols.	Equivalents. Chemical.	Equivalents. Electro-Chemical.	Weights decomposed by one ampere in one hour.
		mgr.		grammes.
Nitrogen	N	14	0·1456	0·5250
Oxygen	O	8	0·0832	0·3
Palladium	Pd	53·2	0·5533	1·9947
Phosphorus	P	31	0·3224	1·1625
Platinum	Pt	98·6	1·0254	3·6975
Potassium	K	39·1	0·4066	1·4637
Selenium	Se	39·8	0·4139	1·4925
Silver	Ag	108	1·1232	4·05
Silicon	Si	14	0·1456	0·5242
Sodium	Na	23	0·2392	0·8625
Strontium	Sr	43·8	0·4555	1·6425
Sulphur	S	16	0·1664	0·6
Tin	Sn	59	0·6136	2·2125
Zinc	Zn	32·7	0·3401	1·2243
Water	HO	9	0·09328	0·03375
Acid, acetic	$C_4H_4O_4$	60	0·6240	2·25
„ nitric (monohydrated)	NO_6H	63	0·6552	2·3625
„ hydrochloric (gas)	HCl	36·5	0·3796	1·3687
„ chromic	CrO_3	50·2	0·5221	1·8825
„ oxalic	$C_2H_2O_8$	90	0·936	3·375
„ phosphoric (hydrated) ..	$PO_5,3HO$	98	1·0192	3·675
„ sulphuric (monohydrated) ..	SO_4H	49	0·5096	1·8375
„ tartaric	$C_8H_5O_{12}$	150	1·56	5·625
Oxides of silver (proto)	AgO	116	1·2064	4·35
„ copper (di)	CuO	39·7	0·4129	1·4864
„ iron (proto)	FeO	36	0·3744	1·35
„ „ (per)	Fe_2O_3	80	0·832	3
„ manganese (di)	MnO_2	43·5	0·4524	1·6312
„ mercury (di)	HgO	108	1·1232	4·05
„ nickel	NiO	37·5	0·39	1·4062
„ lead	PbO_2	119·5	1·2428	4·4812
„ potassium (anhydrous) ..	KO	47·1	0·4898	1·7662
„ sodium	NaO	31	0·3224	1·1625
„ zinc	ZnO	40·5	0·4212	1·5187
Ammonia	NH_4O_1,HO	35	0·3640	1·3125
Argentic nitrate	NO_6Ag	170	1·768	6·375
Chloride of ammonium	NH_4Cl	53·5	0·5564	2·0062
„ iron	FeCl	63·5	0·6604	2·3812
„ „ (per)	Fe_2Cl_3	162·5	1·6900	6·0840
„ nickel	NiCl	65	0·676	2·4375
„ gold (per)	Au_2Cl_3	303·5	3·1564	11·3812
Cyanide of ammonium	$C_2N_2H_4$	44	0·4576	1·65
„ silver	C_2NAg	134	1·3936	5·025
„ potassium	C_2NK	65·1	0·667	2·4412
Sulphate of ammonium	$SO_4N_2H_8$	84	0·8736	3·15
„ copper (anhydrous) ..	SO_4Cu	79·7	0·8289	2·9887
„ „ (crystallised) ..	$SO_4Cu,5HO$	124·7	1·2969	4·6762
„ iron	$SO_4Fe,7HO$	139	1·4456	5·2125
„ nickel	$SO_4Ni,7HO$	140·5	1·4612	5·2687
„ potassium	SO_4K	87·1	0·9058	3·2662
„ sodium	SO_4Na	71	0·7384	2·6625
„ zinc	SO_4Zn	80·7	0·8393	3·0262

FORMATION OF THE PRINCIPAL OXIDES.

Names.	Constituent Elements.	Equivalents.	Heat Developed.	
			Solid State.	Dissolved State.
Water	$H + O$	9	+ 35·2	+ 34·5
Potash	$K + O + HO$	56·1	+ 69·8	+ 82·3
Soda	$Na + O + HO$	40	+ 67·8	+ 77·6
Lithia	$Li + O + HO$	24	..	+ 83·3
Ammonia	$N + H_3 + 2 HO$	35	..	+ 21·0
Calcia	$Ca + O + HO$	37	+ 73·5	+ 75·05
Strontia	$Sr + O + HO$	60·8	+ 74·3	+ 79·1
Magnesia	$Mg + O + HO$	29	+ 71·9	..
Alumina	$Al_2 + O_3 + 3 HO$	78·4	+3×65·3	..
Protoxide of manganese (hydrated)	$Mn + O$	35·5	+ 47·4	..
Dioxide of manganese ,,	$Mn + O_2$	43·5	+ 58·1	..
Chromic acid	$\{Cr_2 + O_3 \text{ hy-} \atop \text{drated} + O_3 \}$	103	+ 3·1	1·4 × 3
Protoxide of iron (**hydrated**)	$Fe + O$	36	+ 34·5	..
Peroxide of iron ,,	$Fe_2 + O_3$	80	+31·9×3	..
Oxide of nickel ,,	$Ni + O$	37·5	+ 30·7	..
,, cobalt ,,	$Co + O$	37·5	+ 32·0	..
,, gold ,,	$Au_2 + O_3$	221	− 5·6	..
,, zinc ,,	$Zn + O + HO$	49·5	+ 41·8	..
,, cadmium ,,	$Cd + O$	64	+ 33·2	..
,, lead ,,	$Pb + O + HO$	120·5	+ 26·7	..
Protoxide of copper	$Cu_2 + O$	71·4	+ 21·0	..
Dioxide of copper (hydrated)	$Cu + O + HO$	48·7	+ 19·0	..
Protoxide of tin ,,	$Sn + O$	67	+ 34·9	..
Dioxide of tin ,,	$Sn + O_2$	75	+ 67·9	..
Protoxide of mercury	$Hg_2 + O$	208	+ 21·1	..
Dioxide of mercury (yellow)	$Hg + O$	108	+ 15·5	..
Oxide of silver	$Ag + O$	116	+ 3·5	..
Protoxide of platinum	$Pt + O$	106·6	+ 7·5	..
Oxide of bismuth	$Bi + O_3$	234	+ 68·9	..
,, antimonious (hydrated)	$Sb + O_3$	146	+ 88·7	..
Antimonic acid	$Sb + O_4$	162	+114·4	..

FORMATION OF THE PRINCIPAL CYANIDES.

Names.	Constituent Elements.	Equivalents.	Heat Developed.	
			Solid State.	Dissolved State.
Hydrocyanic acid	$Cy \text{ gas} + H$	27	7·8 (gas)	+ 13·1
Cyanide of potassium	$Cy + K$	65·1	+ 67·6	+ 64·7
,, sodium	$Cy + Na$	49	+ 60·4	+ 59·9
,, ammonium	$Cy + N + H_4$	44	+ 40·5	+ 36·1
,, zinc	$Cy + Zn$	58·5	+ 29·3	..
,, mercury	$Cy + Hg$	126	+ 11·9	+ 10·4
,, silver	$Cy + Ag$	134	+ 3·6	..
,, mercury and potassium	$Hg Cy + K Cy$	191·1	+ 8·8	+ 6·2
,, silver and potassium	$Ag Cy + K Cy$	199·1	+ 11·2	..

FORMATION OF THE PRINCIPAL CHLORIDES.

Names.	Constituent Elements.	Equivalents.	Heat Developed. Solid State.	Heat Developed. Dissolved State.
Hydrochloric acid	H + Cl	36·5	+ 22·0 (gas)	+ 39·3
Chloride of potassium	K + Cl	74·6	+105·0	+100·8
„ sodium . ..	Na + Cl	58·5	+ 97·3	+ 96·2
„ ammonium ..	N + H₄ + Cl	53·5	+ 76·7	+ 72·7
„ lithium	Li + Cl	42·5	+ 93·5	+101·9
„ calcium	Ca + Cl	55·5	+ 85·1	+ 93·8
„ strontium	Sr + Cl	79·3	+ 92·3	+ 97·8
„ magnesium ..	Mg + Cl	47·5	+ 75·5	+ 93·5
„ aluminium ..	Al₂ + Cl₃	132·9	+ 53·6×3	+ 79·3×3
„ manganese ..	Mn + Cl	63	+ 56·0	+ 64·0
„ iron	Fe + Cl	63·5	+ 41·0	+ 50·0
„ „ (per)	Fe₂ + Cl₃	162·5	+ 32·0×3	+ 42·6×3
„ zinc	Zn + Cl	68	+ 48·6	+ 56·4
„ cadmium	Cd + Cl	91·5	+ 46·6	+ 48·1
„ lead	Pb + Cl	139	+ 42·6	+ 39·2
„ nickel	Ni + Cl	65	+ 37·3	+ 46·8
„ cobalt..	Co + Cl	65	+ 38·2	+ 47·4
„ tin	Sn + Cl	94·5	+ 40·2	+ 40·6
„ „ (bi)	Sn + Cl₂	130	+ 64·6(liq.)	+ 78·7
„ gold	Au₂ + Cl₃	303·5	+ 22·8	+ 27·3
„ copper (proto) ..	Cu₂ + Cl	98·9	+ 35·6	..
„ „ (bi) ..	Cu + Cl	67·2	+ 25·8	+ 31·3
„ mercury (proto)	Hg₂ + Cl	235·5	+ 40·9	..
„ „ (bi) ..	Hg + Cl	135·5	+ 31·4	+ 29·8
„ silver	Ag + Cl	143·5	+ 29·2	..
„ bismuth	Bi + Cl₃	316·5	+ 90·6	..
„ antimony	Sb + Cl₃	228·5	+ 91·4	..
„ „ (per)..	Sb + Cl₅	299·5	+104·9(liq.)	..
Protochloride of platinum and potassium	Pt + Cl + KCl	209·1	+ 22·6	+ 20·9
Bichloride of platinum and potassium	Pt + Cl₂ + KCl	244·6	+ 44·7	+ 42·3

FORMATION OF THE PRINCIPAL BROMIDES (Liquid Bromine).*

Names.	Constituent Elements.	Equivalents.	Heat Developed. Solid State.	Heat Developed. Dissolved State.
Hydrobromic acid	H + Br	81	+ 9·5(gas)	+29·5
Bromide of potassium	K + Br	119·1	+ 96·4	+91·0
„ sodium..	Na + Br	103	+86·7	+86·4
„ ammonium	N + H₄ + Br	98	+67·2	+62·9
„ aluminium	Al₂ + Br₃	267·4	+40·2×3	+69·2×3
„ zinc	Zn + Br	112·5	+39·1	+46·6
„ lead	Pb + Br	183·5	+34·5	+29·5
„ tin	Sn + Br₂	219	+50·7	+59·0
„ copper	Cu + Br	111·7	+17·4	+21·5
„ mercury	Hg + Br	180	+25·9	+24·5
„ silver	Ag + Br	188	+23·7	..
„ gold	Au₂ + Br₃	437	+12·1	+ 8·4

* Gaseous bromine liberates + 4·0 calories when being liquefied.

FORMATION OF THE PRINCIPAL IODIDES (SOLID IODINE).

Names.	Constituent Elements.	Equiva-lents.	Heat Developed.	
			Solid Salt.	Dissolved Salt.
Hydriodic acid	H + I	128	− 6·2 (gas)	+ 13·2
Iodide of potassium	K + I	166·1	+ 80·0	+ 74·7
,, sodium	Na + I	150	+ 68·8	+ 70·1
,, ammonium	N + H$_4$ + I	145	+ 50·6	+ 47·1
,, aluminium	Al$_2$ + I$_3$	408·4	+ 23·4 × 3	+ 53·0 × 3
,, zinc	Zn + I	159·5	+ 24·6	+ 30·3
,, lead	Pb + I	230·5	+ 21·0	..
,, copper	Cu$_2$ + I	190·4	+ 16·5	..
,, mercury (red) ..	Hg + I	227	+ 17·0	..
,, silver	Ag + I	236	+ 14·3	..
,, gold (proto)	Au$_2$ + I	324	− 5·5	..

FORMATION OF THE PRINCIPAL SULPHIDES.

Names.	Constituent Elements.	Equiva-lents.	Heat Developed.	
			Solid Salt.	Dissolved Salt.
Hydrosulphuric acid	H + S	17	+ 2·3 (gas)	+ 4·6
Sulphide of potassium	K + S	55·1	+ 51·1	+ 56·2
,, sodium	Na + S	39	+ 44·2	+ 51·6
,, ammonium	N + H$_4$ + S	34	,,	+ 28·4
,, magnesium	Mg + S	28	+ 39·8	..
,, aluminium	Al$_2$ + S$_3$	75	+ 62·2	..
,, manganese	Mn + S	43·5	+ 22·6	..
,, iron	Fe + S	44	+ 11·9	..
,, zinc	Zn + S	48·5	+ 21·5	..
,, cadmium	Cd + S	72	+ 17·0	..
,, cobalt	Co + S	45·5	+ 10·9	..
,, nickel	Ni + S	45·5	+ 9·7	..
,, lead	Pb + S	119·5	+ 8·9	..
,, copper	Cu + S	47·5	+ 5·1	..
,, mercury	Hg + S	116	+ 9·9	..
,, silver	Ag + S	124	+ 1·5	..

MISCELLANEOUS METALLIC COMPOUNDS.

Names.	Constituent Elements.	Equiva-lents.	Heat Developed.	
			Solid State.	Dissolved State.
Hydruret of platinum	H + mPt	1 + 98·6 m	+ 8·7	..
Another do.	H + nPt	1 + 98·6 n	+ 17·4	..
	Hg$_{24}$ + K	2439·1	+ 34·2	+ 25·7
Amalgam of potassium ..	Hg$_8$ + K	839·1	+ 29·3	dissolved
	Hg$_3$ + K(?)	339·1	+ 16·0	in Hg
	Hg$_{12}$ + Na	1223	+ 21·6	+ 18·8
Amalgam of **sodium**	Hg$_8$ + Na	823	+ 21·1	dissolved
	Hg$_7$ + Na$_2$	746	+ 15·2 × 2	in Hg

FORMATION OF THE PRINCIPAL METALLOID COMPOUNDS.

Names.	Constituent Elements.	Equivalents.	Developed. Free State.	Developed. Dissolved State.
Ozone	$2(O+O_2)$	24×2	$- 14 \cdot 8 \times 2$ (gas)	..
Dioxide of hydrogen (oxigenated water) ..	$H+O_2$	17	..	$+ 23 \cdot 7$
Protoxide of nitrogen ..	$N+O$	22	$- 10 \cdot 3$ (gas)	..
Dioxide of nitrogen	$N+O_2$	30	$- 21 \cdot 6$ (gas)	..
Acid, nitrous	$N+O_3$	38	$- 11 \cdot 1$ (gas)	..
„ hyponitric	$N+O_4$	46	$- 2 \cdot 6$ (gas)	..
„ nitric, anhydrous ..	$N+O_5$	54	$\{ \begin{array}{l} - \ 0 \cdot 6 \text{ (gas)} \\ + \ 1 \cdot 8 \text{ (liq.)} \end{array}$	$\} \begin{array}{l} + 14 \cdot 3 \\ + 14 \cdot 3 \end{array}$
„ „ hydrated ..	$N+O_5+HO$	63	$\{ \begin{array}{l} - \ 0 \cdot 1 \text{ (gas)} \\ + \ 7 \cdot 1 \text{ (liq.)} \end{array}$	$\begin{array}{l} .. \\ - \ 14 \cdot 3 \end{array}$
„ „ 2nd hydrate ..	NO_5H+4HO	99	$+ 5 \cdot 0$ (liq.)	..
„ hyposulphurous ..	S_2+O_2+HO	57	..	$+ 33 \cdot 6$
„ sulphurous	$S+O_2$	32	$+ 34 \cdot 6$ (gas)	$+ 38 \cdot 4$
„ sulphuric anhydrous	$S+O_3$	40	$\{ \begin{array}{l} + 45 \cdot 9 \text{ (gas)} \\ + 51 \cdot 8 \text{ (sol.)} \end{array}$	$\begin{array}{l} .. \\ + 70 \cdot 5 \end{array}$
„ „ monohydrated ..	SO_3+HO	49	$+ 62 \cdot 0$ (liq.)	$+ 70 \cdot 5$
„ „ bihydrated	SO_4H+HO	58	$+ 3 \cdot 1$ (liq.)	..
„ persulphuric	$\{ \begin{array}{l} S_2+O_7 \\ S_2O_8 \text{dissolved} \\ +O \end{array}$	$\} \begin{array}{l} 88 \\ 88 \end{array}$	$\begin{array}{l} .. \\ .. \end{array}$	$\begin{array}{l} +126 \cdot 6 \\ - \ 13 \cdot 8 \end{array}$
„ hypochlorous	$Cl+O$	43·5	$- 7 \cdot 6$ (gas)	$- 2 \cdot 9$
„ chloric hydrated ..	$Cl+O_5+HO$	84·5	..	$- 12$
„ perchloric	$Cl+O_7+HO$	100·5	$- 15 \cdot 4$ (liq.)	$+ \ 4 \cdot 9$
„ „ 2nd hydrate	ClO_5H+2HO	118·5	$\{ \begin{array}{l} + 12 \cdot 6 \text{ (liq.)} \\ + \ 8 \cdot 6 \text{ (sol.)} \end{array}$	$\begin{array}{l} .. \\ .. \end{array}$
„ „ 3rd hydrate	ClO_5H+4HO	136·5	$+ 15 \cdot 0$ (sol.)	..
Oxide of carbon	$C+O$	14	$+ 14 \cdot 4$ (gas)	..
Carbonic acid	$C+O_2$	22	$+ 48 \cdot 5$	..
Sulphide of carbon	$C+S_2$	38	$\{ \begin{array}{l} - \ 9 \cdot 05 \text{ (gas)} \\ - \ 5 \cdot 7 \text{ (liq.)} \end{array}$	$\begin{array}{l} .. \\ .. \end{array}$
Silicic acid	$Si+O_2$	30	$109 \cdot 6$ (sol.)	$103 \cdot 7$
Chloride of silicon	$Si+Cl_2$	85	$\{ \begin{array}{l} - 75 \cdot 6 \text{ (gas)} \\ - 78 \cdot 8 \text{ (gas)} \end{array}$	$\begin{array}{l} .. \\ .. \end{array}$
Organic Compounds.				
Acid, acetic	$C_4+H_4+O_4$	60	$+129 \cdot 1$ (sol.)	$+127 \cdot 0$
„ oxalic	$C_4+H_2+O_8$	90	$+197$ (sol.)	$+194 \cdot 7$
„ tartaric	$C_8+H_6+O_{12}$	150	$+372$ (sol.)	$+368 \cdot 7$
„ citric	$C_{12}+H_8+O_{14}$	192	$+354$ (sol.)	..

FORMATION OF THE AMMONIACAL COMPOUNDS (Acid and base gaseous, unless otherwise specified).

Names.	Constituent Elements.	Heat Developed.
Ammonia	$N + H_3$	$\begin{cases} + \ 12\cdot2 \ (\text{gas}) \\ + \ 21\cdot0 \ (\text{dissolved}) \end{cases}$
Hydrochlorate	$HCl + NH_3$	$+ \ 42\cdot5$
Hydrobromate	$HBr + NH_3$	$+ \ 45\cdot6$
Hydriodate	$HI + NH_3$	$+ \ 44\cdot2$
Hydrocyanate	$HCy + NH_3$	$+ \ 20\cdot5$
Hydrosulphate	$H_2S_2 + NH_3$	$+ \ 23$
Sulphate	$SO_4H + NH_3$	$+ \ 25\cdot8 \left(\begin{smallmatrix} SO_4H \\ \text{dissolved} \end{smallmatrix}\right)$
Acetate	$C_4H_4O_4 + NH_3$	$+ \ 26\cdot0$
Nitrate	$NO_6H + NH_3$	$+ \ 41\cdot9$
Oxalate *	$\frac{1}{2} \ C_4H_2O_6 + NH_3$	$+ \ 24\cdot4$
Bicarbonate	$C_2O_4 + H_2O_2 + NH_3$	$+ \ 30\cdot4$

* In the oxalate, the acid is taken in the hydrated state and solid.

FORMATION OF THE PRINCIPAL SALTS DISSOLVED OR PRECIPITATED BY MEANS OF DISSOLVED ACIDS AT ABOUT 15°.

Names of the Bases.	Symbols.	Nitrates, NO_6H	Acetates, $C_4H_4O_4$	Oxalates, $\frac{1}{2}C_4H_2O_6$	Sulphates, SO_4H	Carbonates, CO_2	Chlorides, HCL
Soda	NaO	13·7	13·3	14·3	15·85	10·2	13·7
Potash	KO	13·8	13·3	14·3	15·7	10·1	13·7
Ammonia	NH_3	12·5	12·0	12·7	14·5	5·3	12·45
Calcia	CaO	13·9	13·4	18·5(pr.)	15·6	9·8(pr.)	14·0
Baryta	BaO	13·9	13·4	16·7	18·4(pr.)	11·1	13·85
Strontia	SrO	13·9	13·3	17·6	15·4	10·5	14·0
Magnesia	MgO	13·8	..	..	15·6	9·0	13·8
Protoxide of manganese ..	MnO	11·7	11·3	14·3	13·5	6·8	11·8
Protoxide of iron	FeO	..	9·9	..	12·5	5·0	10·7
Oxide of nickel ..	NiO	..	..	..	13·1	..	11·3
„ cobalt ..	CoO	..	..	..	13·3	..	10·6
„ cadmium	CdO	10·1	..	..	11·9	..	10·1
„ zinc ..	ZnO	9·8	8·9	12·5	11·7	5·5	9·8
„ lead ..	PbO dilute	7·7	6·5	12·8	10·7(pr.)	6·7	7·7
„ copper ..	CuO	7·5	6·2	..	9·2	2·4	7·5
„ mercury	HgO(prec.)	..	3·0	7·0	..	..	9·45
„ silver ..	AgO(prec.)	5·2	4·7	12·9	7·2	6·9	..
Alumina, hydrated ..	$\frac{1}{2} Al_2O_3$	..	..	..	10·5	..	9·3
Sesquioxide of iron, hydrated	$\frac{1}{2} Fe_2O_3$	5·9	4·5	..	5·7	..	5·9
Sesquioxide of chromium ..	$\frac{1}{2} Cr_2O_3$	..	..	..	8·2	..	6·9

FORMATION OF THE DISSOLVED SALTS.   Various Acids at about 15°.

Names of the Acids.	Symbols.	Names of the Bases.	Symbols.	Heat Developed.
				deg. C.
Acid hydrofluoric	HF dissolved	Soda dilute	NaO	+ 16·3
„ nitrous	NO_3HO dilute	Ammonia	NH_3 dilute	+ 8·8
„ perchloric	ClO_7, HO dilute	Soda dilute	NaO	+ 14·2
		Baryta dilute	BaO	+ 14·5
		Ammonia	NH_3 dilute	+ 12·9
„ hypochlorous ..	ClO, HO dilute	Soda dilute	NaO	+ 9·5
		Baryta dilute	BaO	+ 9·8
„ hyposulphurous ..	S_2O_2HO dilute	Soda dilute	NaO	+ 13·5
„ sulphurous	SO_2 dissolved	Potash dilute	KO	+ 15·9
„ „	2 SO_2 dissolved	„	KO	+ 17·9
„ hypophosphorous	PO, 3 HO dilute	Soda dilute	NaO	+ 15·2
„ phosphorous ..	PO_3, 3 HO dilute	„	NaO	+ 14·8
		„	2 NaO	+ 28·4
„ phosphoric mono-hydrated	PO_5, HO dilute	„	NaO	+ 14·5
„ phosphoric bi-hydrated	PO_5, 2 HO dilute	„	NaO	+ 14·8
		„	2 NaO	+ 26·4
„ arsenic	AsO_5 dissolved	„	NaO	+ 15·0
„ „	„	„	2 NaO	+ 27·6
„ „	„	„	3 NaO	+ 35·9
„ arsenious	AsO_3 dissolved	„	NaO	+ 13·8
„ „	„	„	2 NaO	+ 15·1
„ silicic	SiO_2 gelatinous	„	½ NaO	+ 2·15
„ stannic	SnO_2 gelatinous	?	NaO	+ 4·8
„ chromic	CrO_3 dilute	„	NaO	+ 12·4
„ tartaric	$C_8H_4O_{12}$ dissolved	„	NaO	+ 12·7
		„	2 NaO	+ 12·6

MOTIVE POWER required for the ELECTROLYTICAL SEPARATION OF VARIOUS METALS FROM THEIR SALINE SOLUTIONS.

Metal.	Designation of the Electrolyte.	Heat of Combination taken in relation to 1 Kilogramme of the Metal.	Motive Power required for the Decomposition of 1 Kilogramme of the Electrolyte.		Motive Power required for the deposition of 1 Kilogramme of the Metal.		Weight of the Metal deposited per Horse-power per hour.
			Kilogramme-metres.	Horse-power per hour.	Kilogramme-metres.	Horse-power per hour.	
		cal.					kil.
Copper ..	Sulphate of copper ..	886	149,800	0·55	375,700	1·39	0·719
,, ..	Chloride of copper ..	990	197,600	0·73	419,800	1·55	0·645
Silver ..	Double cyanide of silver and potassium	33	11,470	0·04	14,220	0·05	18·900
,, ..	Nitrate of silver dissolved in ammonia	81	21,630	0·08	34,340	0·13	7·700
Gold ..	Chloride of gold (with cyanide of potassium)	139	38,100	0·14	58,700	0·22	4·600
Nickel ..	Double sulphate of nickel and ammonium	1,484	239,600	0·89	629,400	2·33	0·429
,, ..	Double chloride of nickel and ammonium	1,586	305,300	1·13	671,700	2·49	0·402
Tin	Protochloride of tin ..	688	182,100	0·67	291,400	1·08	0·926
Lead ..	Acetate of lead	320	86,600	0·32	136,000	0·51	1·985
,, ..	Nitrate of lead	332	88,100	0·33	141,000	0·52	1·915
Zinc ..	Double oxalate of zinc and ammonium ..	740	133,800	0·50	314,400	1·16	0·858
,, ..	Double cyanide of zinc and potassium ..	896	211,600	0·78	379,800	1·41	0·711
Mercury ..	Chloride of mercury ..	298	93,200	0·35	126,800	0·47	2·129
Aluminium	Chloride of aluminium and ammonium ..	8,680	758,700	2·81	3,679,700	13·60	0·073
Magnesium	Chloride of magnesium and ammonium ..	7,665	831,100	3·08	3,249,600	12·00	0·083
Potassium*	Sulphate of potassium	2,512	477,600	1·77	1,065,000	3·95	0·253
,,	Chloride of potassium	2,584	573,600	2·13	1,094,000	4·05	0·247
Sodium† ..	Sulphate of sodium ..	4,063	558,400	2·07	1,725,500	6·39	0·156
,, ..	Chloride of sodium ..	4,183	697,300	2·58	1,773,200	6·57	0·152
Antimony	Trichloride of antimony	749	169,600	0·63	317,200	1·17	0·851
Iron ..	Double chloride of iron and ammonium	1,785	333,800	1·24	754,400	2·79	0·358
,, ..	Double oxalate of iron and ammonium ..	584	96,200	0·36	247,300	0·92	1·091
Platinum	Chloride of platinum (with cyanide of K)	209	65,500	2·43	88,616	0·33	3·050
Hydrogen	Water acidulated with sulphuric acid ..	34,500	1,625,200	6·02	14,626,800	54·18	0·0184

* The metal is supposed to be in a free state.
† Same observation.

LIMITS OF ELECTROLYSIS (SUMMARY OF M. BERTHELOT'S EXPERIMENTS).

Designation of Substances and Symbols.	Separated Substances. To the Positive Pole.	To the Negative Pole.	Heat absorbed.	Maximum No. of Calories of the Battery. Which still cause a Decomposition.	Which do not cause a Decomposition.	Observations.
Sulphate of potassium KSO_4 ..	1. SO_4H dilute $+ O$	KO dil. $+ H$	cal. $-50\cdot2$	$51\cdot5$	46	In these experiments, electrolysis has been studied with an apparent object; the electrodes are short platinum wires unless otherwise specified.
	2. Positive pole of copper: CuO, SO_3 dilute	KO dilute $+ H$	-22	$24\cdot5$	19	
	3. Positive pole of zinc: ZnO, SO_3 dil.	KO dilute $+ H$	$+3$	..	..	The decomposition occurs without the battery.
	4. Positive pole of platinum: SO_4H dilute $+ O$	Negative pole of mercury: K (amalg.)	$-72\cdot3$	$73\cdot5$	68	The electrolysis occurs at the inferior limit but without the formation of an amalgam
Sulphate of magnesium	SO_4H dilute $+ O$ {	MgO dilute $+ H$	-50	54	46	Same decomposition as for KSO_4.
Sulphate of zinc: $ZnSO_4$ ..	1. SO_4 dilute $+ O$	Zn	$-53\cdot5$	57	$51\cdot5$	Hydrogen is at first liberated.
	2. Positive pole of copper: $CuOSO_3$ dilute	Zn	$-24\cdot5$	$24\cdot5$	$24\cdot5$	Is explained by some differences of concentration.
Sulphate of cadmium	SO_4H dilute $+ O$	Cd	$-45\cdot1$	46	38	
Sulphate of copper ..	SO_4H dilute $+ O$	Cu	$-28\cdot2$	$32\cdot5$	27	
Chloride of potassium KCl ..	Cl dissolved with the formation of various compounds	KO dilute $+ H$ {	-46 to -47	46	$40\cdot5$	Haloid salts; decomposition analogous to that of sulphate of potassium; this remark applies to the decomposition of all the haloid salts in general.
Bromide of potassium KBr ..	Br dissolved with the formation of a perbromide	KO dilute $+ H$ }	$-41\cdot5$	40	40	
Fluoride of potassium KF ..	FH dilute $+ O$ {	KO dilute $+ H$ }	-51	>51	51	
Iodide of potassium KI	I	KO dilute $+ H$ }	-27	27	$24\cdot5$	

SALTS THE METALS OF WHICH OFFER SEVERAL DEGREES OF OXIDATION.

Designation of Substances and Symbols.	Separated Substances. To the Positive Pole.	To the Negative Pole.	Heat absorbed.	Which still cause a Decomposition.	Which do not cause a Decomposition.	Observations.
Ferrous sulphate $FeSO_4$	1. SO_4H dilute $+ 2SO_3, Fe_2O_3$ dil.	Fe	-32	$32\cdot5$	27	In this 2nd case the production of the ferric salt continues.
	2. SO_4H dilute $+ O$	$FeO + H$	-47	$57\cdot0$	49	
Manganous sulphate $MnSO_4$ ..	1. $MnO_2 + SO_4H$	H	-37	38	27	In this 2nd case MnO_2 continues to be deposited at the anode.
	2. SO_4H dilute $+ O$	Mn	$-60\cdot9$	$62\cdot5$	57	

CASE OF THE POLARISATION OF THE ELECTRODES.

Designation of Substances and Symbols.	Separated Substances. To the Positive Pole.	To the Negative Pole.	Heat absorbed.	Which still cause a Decomposition.	Which do not cause a Decomposition.	Observations.
Nitrate of potassium KNO_4 ..	1. NO_6H dilute $+ O$	KO dil. $+ H$ (H absorbed by the acid)	-26 to -16	64	27	With all these substances, the reaction which begins is stopped by the momentary contact of the electrodes. The inversion of the current or of the poles causes its resumption. Sometimes the escaping of gases spontaneously ceases or is not resumed after the contact of the electrodes.
	2	..	..	57	38	
Sulphate of ammonium NH_4SO_4 ..	SO_4H dilute $+ O$	NH_4 (with oxidation of the ammonia)	-37 to -35	38	$24\cdot5$	
Acetate of sodium $C_4H_3NaO_4$	$C_2H_3g + C_2O_4$ dissolved $+ HO$ {	NaO dilute $+ H$	-38	38	35	

SECTION II.

CHAPTER V.

BATTERIES.

Hydro-electric Batteries in Electro-chemistry—Becquerel Cell—Daniell Cell
—Simple Bath—Balloon Cell—Thomson Cell—Callaud Cell—Meidinger
Cell—Bunsen Cell—D'Arsonval Cell—Duchemin Cell—Leclanché Cell—
Oxide of Copper Cell—Smee Cell—Noé Thermo-electric Battery—Clamond
Thermo-electric Battery.

HYDRO-ELECTRIC BATTERIES IN ELECTRO-CHEMISTRY.—The
hydro-electric batteries mostly used in electric operations are
those of Daniell, Bunsen, and like inventors. For thin depositions, and especially when the work is intermittent, Leclanché,
Lalande and Chaperon's cells can be used, as also generally all
cells having the property of giving a regular current for a long
time without wasting when not at work.

We will briefly describe the cells in use, and draw attention
to the alterations which could be made in them in order to make
them more economical or less offensive in industrial practice.

BECQUEREL CELL.—To Becquerel is due the first partitioned cell with two liquids and two metals. This cell consisted of a glass vessel partitioned in two compartments by
means of a sheet of gold-beater's skin; in one compartment
were placed a solution of nitrate of copper and a strip of copper
sheet, and in the other a solution of nitrate of zinc and a strip
of zinc sheet. The Becquerel cell is almost identical with that
which has been called the Daniell cell, and which is considered

to be the prototype of the two-liquid cell. The latter is, however, more constant.

The object of the two compartments is to prevent the polarisation by interposing, between the two electrodes, a liquid which does not attack the positive metal, while at the same time it supplies sufficient quantities of oxygen for consuming the hydrogen as fast as the latter is becoming deposited on the negative metal.

DANIELL CELL.—The Daniell cell, as most generally in use in France, consists of an external vessel made of earthenware, a hollow cylinder of zinc, an internal porous china vessel, and a central strip of copper. In the internal porous vessel is a saturated solution of sulphate of copper, and in the external vessel dilute sulphuric acid. The copper electrode dips into the sulphate, and the zinc one into the acidulated water.

Fig. 1 is an illustration of the Daniell cell as it is in actual use. The zinc is dissolved by oxidisation, and forms

Fig. 1.

sulphate of zinc; the hydrogen which results from this reaction, passing through the porous vessel, substitutes itself in the sulphate for an equal quantity of copper, the latter being deposited on the negative electrode.

After a certain time of working, all the sulphuric acid of the external vessel is converted into sulphate of zinc; the action, nevertheless, continues without any change in the electromotive force. The chemical reactions are then reduced to the substitution of the zinc for the copper in the sulphate of copper, and to the progressive conversion of the sulphate of copper into sulphate of zinc. It is even unnecessary to use dilute sulphuric acid when charging the cell, pure water being sufficient; the action is weaker at starting, but the radical SO_4 rapidly gets through the partition, and is converted into sulphate of zinc by the action of the zinc.

The electromotive force of a Daniell cell differs slightly from one volt, and it is constant enough to be taken as a unit in ordinary practice.

The capital defect—the only defect, we should say—which can be set against the Daniell cell, is that it consumes nearly as much sulphate of copper and zinc in open circuit, when it does not give off any useful work, as in closed circuit when it is really used as a generator of electricity.

BALLOON CELL.—In order to avoid having to maintain the sulphate of copper cell, some makers—Bréguet particularly—have arranged a glass flask, resting in an inverted position, on the vessel of the cell (Fig. 2). This flask contains one kilogramme of crystals of sulphate of copper, and is filled up with water; it is closed by means of a cork stopper. The cork stopper is pierced with a hole, in which is fitted a guttapercha tube, reaching the liquid contained in the porous vessel. The solution of sulphate of copper being the denser according to its degree of concentration, the bottom portion of the liquid is always kept in a state of saturation. As it becomes weaker, it is replaced by the saturated solution which descends from the flask. The external vessel itself is closed by a wooden ring, so that the evaporation is very slight, creeping salts are formed with difficulty, and the liquid lasts a very long time.

FIG. 2.

WILLIAM THOMSON CELL.—The Daniell cell which has been modified by Sir William Thomson in order to obtain elements of a small internal resistance, is composed of wooden trays lined with a sheet of lead; on this lead rests the copper strip which constitutes the negative electrode. At the four corners of these trays are placed small wooden blocks which support a zinc grating, the latter being the positive electrode. The cells are arranged in a column, as is the case with a voltaic battery, this arrangement producing a powerful battery in a comparatively limited space. The zinc is surrounded by a sheet of parchment which acts as a porous vessel, and prevents the mixture of the liquids. The battery is charged with a solution of sulphate of zinc, to which is added the sulphate of copper in small crystals and placed as regularly as possible all around and at the bottom of the tray.

CALLAUD CELL.—The Callaud cell (Fig. 3) is a Daniell cell without a porous jar, and in which the separation of the liquids is obtained through their difference of density. It is

FIG. 3.

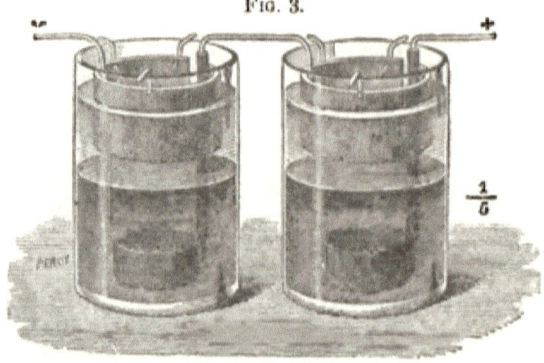

composed of a glass cylindrical jar, at the bottom of which a ring made of a thin strip of copper is laid, and a cylinder of zinc is suspended from the brim by means of three small hooks. The solution of sulphate of copper being denser than that of sulphate of zinc, keeps at the bottom of the jar, so that the two electrodes are always dipping into their respective liquids without any possibility of getting mixed.

The copper rod which is used for suspending the cylinder of copper is covered with guttapercha, without which precaution experience has shown that it would in time part at the line of separation of the two liquids. The addition of sulphate of copper, when necessary, is effected by means of a siphon. This cell is more particularly employed in the United States.

MEIDINGER CELL.—The Meidinger cell, constructed on the same principles as the preceding one, is largely used in Russia, Austria, and Germany. We represent it (Fig. 4) under its best known form. The external vessel is larger at the top than at the bottom. The sulphate of copper and the negative strip are contained in a small conical jar placed at the bottom of the vessel. An inverted flask, the neck of which reaches into the above-mentioned conical jar, is· filled with crystals of sulphate of copper, which, owing to the large capacity reserved for the sulphate of zinc, allows the battery to be used for a long time without rendering it necessary to take the cells to pieces and to put them together again.

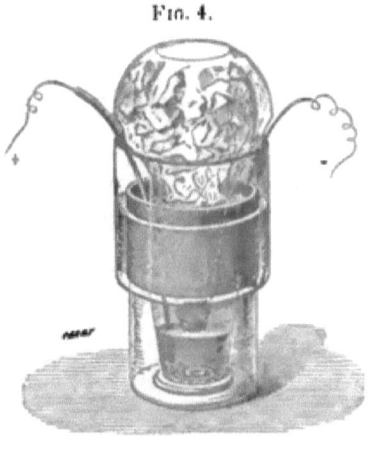

FIG. 4.

Under this perfected shape, M. Meidinger's cell may be considered as one of the best forms of Daniell cell in use, and as one of those which consume the least sulphate of copper for the production of a given quantity of electricity; but its resistance is comparatively great.

BUNSEN CELL.—In his original cell M. Bunsen made use of a cylinder of carbon dipping into the nitric acid contained in an external jar, and a strip of zinc dipping in the acidulated water contained in an internal porous jar. Without making any change in the liquids, or in the electrodes, M. Archereau placed the carbon in the centre and the zinc outside, and it is under that form (Fig. 5) that the Bunsen cell is used in France.* This battery is the most generally used in industry, notwithstanding the acid smell which it emits. It owes its sustained success to its great electromotive force and its small resistance.

---

* We are not writing a treatise on electric batteries, or we should not fail to mention the Grove cell as having preceded that of Bunsen in its general arrangement.

The Bunsen cell is composed of an external glazed earthenware jar, a strip of zinc of cylindrical shape and well amalgamated, an internal porous jar, and a prism of carbon placed

FIG. 5.

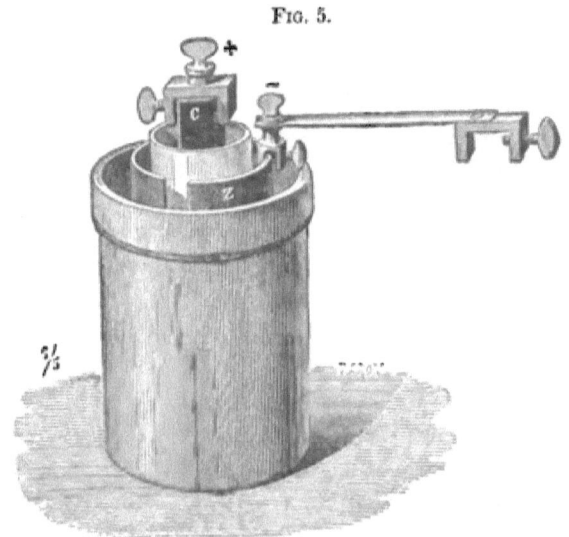

inside the latter. The external jar contains water acidulated
by means of sulphuric acid which has been manufactured with
sulphur and not with pyrites. The porous jar contains nitric
acid.

To prepare the battery, all the junctions are carefully
inspected and cleaned; then the acidulated water is poured
into all the earthenware jars, and the zincs, well amalgamed,
placed in the liquid; the porous jars containing the nitric acid
and the carbon prisms are next placed in the external jars, and
the terminals are metallically joined, the positive ones together
and the negative ones together if it is desired to couple the
cells in quantity or intensity, or the positive of one to the
negative of the following one, and so on, for coupling in tension.

D'ARSONVAL BATTERY.—M. d'Arsonval has succeeded in
greatly reducing the cost of electricity, by increasing the production in a great measure, and in getting rid of a certain
amount of inconvenience which is attached to the use of the

Bunsen battery. He has **preserved the** general arrangement of the latter.

Instead **of** a prism of carbon, **M. d'Arsonval** introduces, in the porous jar, a bundle of cylindrical carbon pencils one centimetre in diameter. These pencils are united together by dipping one end of the bundle for a few minutes **in** boiling paraffin; when cold, this end is coated with a galvanic **deposit of copper,** and finally covered with a ring of printing-type **metal cast** upon it. The depolarising surface is thereby greatly **increased** at the same time that the internal **resistance of the cell is** reduced.

The zinc preserves its form and its position, but the **liquid in** which it is immersed is modified. Instead of dilute sulphuric acid, it is a mixture composed **of** hydrochloric and sulphuric **acids,** the volume of both being in the proportion of $\frac{1}{20}$. The sulphuric acid is not, in this case, obtained **from sulphur, as is** the case with the ordinary Bunsen cell, **but from pyrites; to** purify it, one-sixth of **an** ounce of colza **oil is poured into the** sulphuric acid for each quart of the **latter. The oil is decom-**posed, sulphoglyceric **acid** is formed, and the impurities **are** precipitated in a state of insoluble soaps.

The presence of fat acids facilitates the amalgamation, and the zinc is scarcely attacked when the circuit is opened.

In order to prevent the zinc being attacked by the filtration of the nitric acid through the **porous** jar, M. d'Arsonval mixes **a** small quantity of sulphate of soda in the acidulated **water.**

The depolarising **solution in which** the bundles **of** carbons are immersed is of the following **composition:—**

Nitric acid	1 volume.
Hydrochloric acid	1 „
Acidulated water **per** $\frac{1}{20}$ **of** HO, SO_2	2 volumes.

Thus prepared, one cell 8 inches high **can** easily give 25 amperes per second with an electromotive force of 2 volts, which corresponds to an electrical energy of 5 kilogrammetres.

In order to avoid **any** smell, M. d'Arsonval combines a cell with broken **carbon and a** continuous flow **in** the porous cell of a liquid composed of equal volumes of water saturated when cold with bichromate of potash and hydrochloric acid.

F

We believe M. d'Arsonval's cells to be excellent, and particularly the first one, and should like to see their use become popular. Perhaps it might be possible to replace the crown of carbon pencils by the ordinary prismatic carbon wrapped in cloth carbonised by the Caron process. The depolarising surface would thus be increased, and the carbon part would be stronger and necessitate less frequent cleanings. The rutilant vapours which are emitted from every Bunsen cell can be suppressed without altering the depolarising agent and without complicating the cell by a system of a continuous flow of liquid. This is effected by the very simple and efficacious means of covering the surface of the porous cell with a layer, about 1 inch thick, of ordinary colza oil. M. Rousse, Professor of Physics to the Saint Etienne "lycée," discovered it, and M. Schreurs experimented on it with success in M. Gramme's laboratory.

DUCHEMIN CELL.—This cell, which is used in M. Oudry's works at Auteuil for the process of electro-gilding, is a Bunsen cell in which the nitric acid has been replaced by a solution of sesquichloride of iron. The sesquichloride of iron is reduced by the hydrogen, yielding monochloride of iron and hydrochloric acid, which remains in solution. In this way the polarisation is destroyed as in the Bunsen cell, and there is no emission of gas. When the battery has been working for a certain length of time, and the salts of iron have been brought from a maximum to a minimum, it is sufficient in order to put the cell in working order to send a flow of chlorine through the solution of the salts of iron or to boil them with a few drops of nitric acid. A certain economy is thus obtained, but the manipulation of acid, however small, always acts prejudicially against the propagation of any system of battery.

LECLANCHÉ CELL.—The Leclanché cell (Fig. 6) is composed of a square glass vessel, contracted at its top, a positive electrode consisting of a simple pencil of zinc, and a negative electrode consisting of a central carbon plate and two agglomerate plates. The zinc pencil, the carbon, and the agglomerate plates are kept together by means of indiarubber rings. The liquid contained in the glass cell is a solution of sal-ammoniac.

The agglomerate plates are composed of 40 parts of binoxide

of manganese, 52 of carbon, 3 of bisulphate of potash, 5 of gum lac rosin acting as a cement to unite the two other substances. This mixture is heated to 100 degrees Centigrade and then submitted to a pressure of 4500 lb. to the square inch.

Fig. 6.

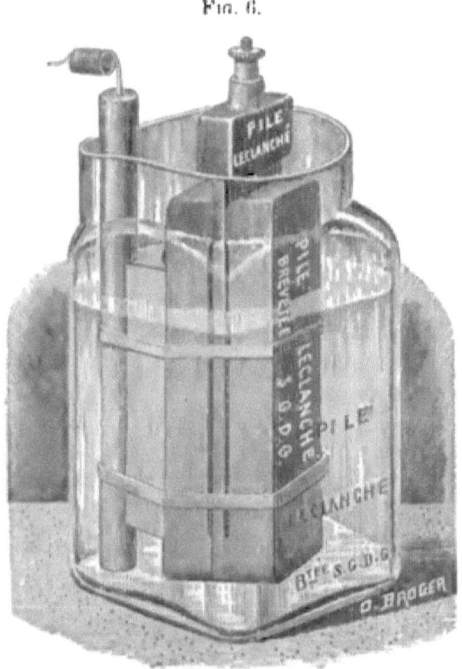

Leclanché Cell.

The bisulphate of potash facilitates the dissolution of the chloric oxide, which in time penetrates through the pores of the mixture.

When the circuit is closed, the current decomposes the solution of sal-ammoniac, and a chloric oxide of zinc soluble in the surrounding liquor is formed; the hydrogen and the ammonia go to the negative electrode, where they effect the reduction of the binoxide of manganese, thereby suppressing all polarisation; the hydrogen becoming oxidised forms water, and the binoxide is reduced to sesquioxide. When preparing the liquid it is necessary to use sal-ammoniac free from metallic salts, and especially of lead, as the latter, depositing itself in a metallic state in a black spongy mass upon the zinc, would soon destroy it.

The Leclanché cell is largely used for telegraphic purposes, but it can be used with advantage in electrolytic operations when the baths have a great resistance, as the work is intermittent and the number of amperes small. It is possessed of a series of properties which make it quite advantageous in a large number of cases. For instance, the zinc not being attacked by sal-ammoniac, no chemical action takes place when the circuit is open; a cell of dimensions equal to a Daniell has a less resistance than the latter; it does not contain any poisonous substance; does not emit any acid vapours or appreciable smells; it resists intense cold without freezing, and consequently without any interruption in its action, &c.

OXIDE OF COPPER CELL.—In the same order of ideas we may mention the oxide of copper cell, the invention of Messrs.

FIG. 7.

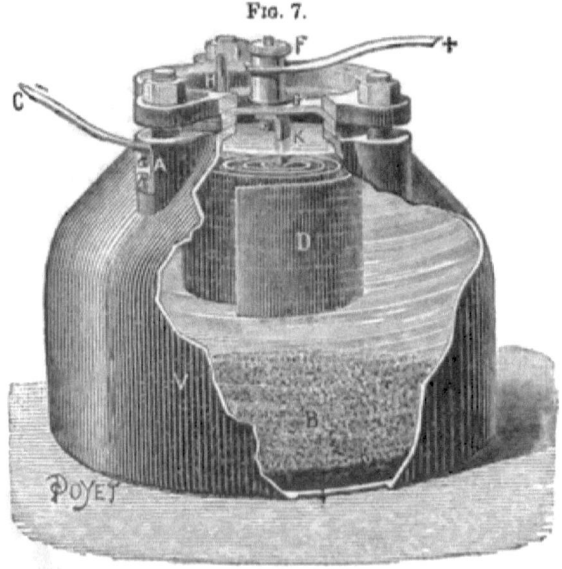

Oxide of Copper Cell.

De Lalande and Chaperon (Fig. 7). It is essentially composed of an external vessel V of thin cast iron, having the shape of a large ink-well and constituting the negative electrode, and a zinc D coiled in a spiral **shape** and suspended from the ebonite cover G. The wires C and F are the electrodes, C being attached to the boss A, and **F to** a terminal which communicates with the positive electrode.

The external cell V contains **oxide of copper B on the top** of which **is** poured a depolarising solution **of caustic** potash **of** 30 or 40 **per 100.**

The cover **is hermetically sealed by means of three flanges** screwed down by **as many nuts.**

When the circuit **is closed** the water is decomposed, **the** oxygen forming with the **zinc** oxide of zinc which combines with the potash to **form an** excessively soluble zincate **of** potash; the hydrogen reduces the oxide of copper **to a** metallic state.

In open circuit **the** materials are **not** attacked, therefore **there** is no reaction **and** consequently **no** expenditure.

The electromotive force varies **between 0·8** and **0·9** volt, and it remains perfectly constant for a long time.

Smee Cell.—The Smee cell is simple and economical, but its depolarisation is less complete than that of cells with two liquids. It is composed **of a** vessel containing **dilute** acid, a positive electrode of amalgamated zinc, **and a negative electrode** of platinised **silver; the bubbles of hydrogen escaping more** freely on such a deposited surface of grey pulverulent rough platinum than on **a polished metallic surface.**

In order to increase **the constancy** of this cell, a small quantity of platinum **chloride is** often **added** to the liquid. Some makers, instead of platinised silver plates, **use** copper plates, on which are successively deposited a layer of galvanic copper, a layer of granulous silver, and a layer of pulverulent platinum.

The Smee cell is extensively **used** in England and in America in small nickeling **and** electrotyping works. It is particularly applicable for operations in which the current must be energetic at the beginning and may be weaker afterwards.

In France it has been almost forgotten, notwithstanding the economy which it obtains over the generality of other systems.

NOÉ THERMO-ELECTRIC BATTERY.—The Noé battery (Fig. 8) is composed of German silver and an antimonious alloy. The active solder is not directly heated; it is inclosed in a brass cap from the centre of which emerges a rod of red copper, one end of which terminates in a point. The said point is heated by the jet of a gas-burner.

FIG. 8.

Noé Thermo-Electric Battery.

The cold soldering is made with tin. In order to facilitate the cooling, a few thin sheets of brass of great surface are soldered together; the surrounding air maintains them at a low temperature. This cell or element has an electromotive force of about $\frac{1}{15}$ volt and an internal resistance of $\frac{1}{30}$ ohm.

A battery of 25 elements decomposes water.

In Austria, the goldsmiths frequently use a Noé battery of 40 elements and heated by gas, for their operations of gilding, silvering, and nickeling. Its capacity is not great, but the apparatus is a simple one. In the various works where we have seen it in operation, we have been told that its duration was to a certain extent considerable.

CLAMOND THERMO-ELECTRIC BATTERY.—M. Clamond has equally adopted the zinc-antimony alloy recommended by

Marcus, but he has selected, as a second metal, iron in preference to copper, as the former is more lasting.

We illustrate (Fig. 9) the model of a Clamond battery which we had occasion to use in 1875 for actuating a small Gramme machine.

This battery consists of 10 elements circularly arranged, and therefore possesses ten bars of Marcus alloy connected

FIG. 9.

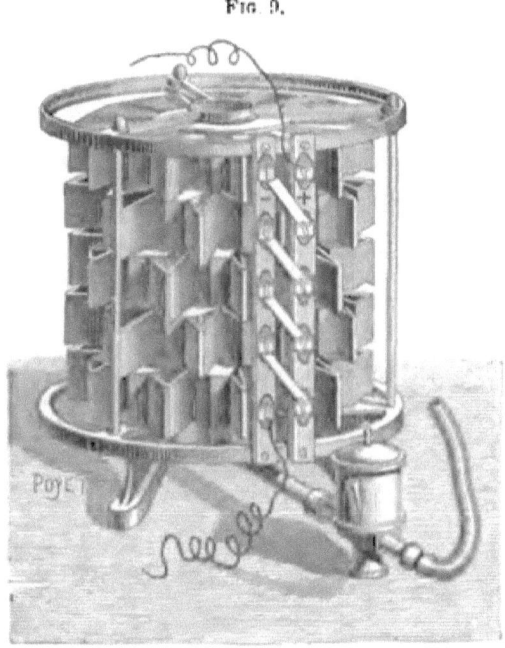

Clamond Thermo-Electric Battery.

together by means of thin sheets of iron offering externally a large surface to the cooling action of the atmosphere. The even soldered joints are at the centre, and the odd ones outside the apparatus.

Such a battery, for example, composed of 5 crowns of 10 couples each, regularly deposits 20 grammes of copper per hour, with an expenditure of 170 litres of gas (6 cubic feet), which corresponds, as we have seen, to an intensity of about 17 amperes. If we admit a total electromotive force of half a volt, which is certainly exaggerated, the work obtained will be found inferior to 1 kilogrammetre. The heat generated by 1 cubic metre of gas would develop, through the intermediary of a Clamond battery, a maximum power of 5 kilogrammetres, whereas the same quantity of gas consumed in an improved "Otto" gas engine develops over 75 kilogrammetres.

CHAPTER VI.

DYNAMO-ELECTRIC MACHINES.

General Characteristics of Dynamo-Electric Machines—Gramme Machine—
Siemens Machine—Schuckert Machine—Mather Machine—Wilde Machine—Weston Machine—Elmore Machine—Superiority of Dynamo-Electric Machines over Batteries.

GENERAL CHARACTERISTICS OF DYNAMO-ELECTRIC MACHINES.
—The use of batteries has, in a large number of electroplating
and electro-metallurgic works, been superseded by that of
dynamo-electric machines. These are much simpler of installation, they occupy less room, generate electric currents
cheaply, and do not give off any smell.

These apparatuses are all based on Faraday's experiments.
Faraday was the first to demonstrate that electric currents
could be produced by means of magnets and conductors having
a relative motion to each other. It was already known, from
Ampère and Arago's discoveries, that a bar of iron could be
magnetised by an electric current. Faraday reciprocally
demonstrated that a magnet could generate an electric current.

A dynamo-electric machine is any machine capable of
giving rise to an electric current in a closed metallic circuit,
by the fact of the displacement of a portion of the said circuit
in the field of action of a magnet.*

The displacement is relative in this respect, that if the
magnet is moving and the wire stationary, there is equally
production of electricity.

The current so generated resulting only from the displacement of a metallic wire in the vicinity of a magnet, it was
perceived that by increasing the magnetic action, the length of

* This definition, as precise as it is short, is due to Antoine Bréguet.

the wire and the relative speed of the two organs, the effects would become more powerful. This is indeed what experience first, and then theory, completely demonstrated. Instead of a magnet, an electro-magnet excited by an external source, or more frequently by the machine itself, is generally used.

The magnet or electro-magnet which influences the conducting wire is called *inductor*, and the portion of the conductor which moves in the magnetic field is called *induced*.

The first machine constructed with a view of realising Faraday's conception, generated currents alternately positive and negative which had to be redressed by means of special devices called commutators before they could be used for electroplating purposes. These commutators wore out rapidly, and gave rise to large sparks, so that these primitive machines did not offer sufficient advantages over batteries to supersede the latter. In reality they met with no success, and were employed on a very limited scale only.

M. Zenobe Gramme produced the first practical continuous-current machine. His works revolutionised the old electrical industries, and gave rise to some new ones. It is therefore only just to begin our review of the various machines used in electrolysis with that of this celebrated inventor.

GRAMME MACHINE.—The Gramme machine is composed of an iron wire ring on which are wound copper wires, and of a cylinder made of a series of blades or segments attached to the said copper wires; the whole mounted on a spindle and revolving inside any given magnetic field.

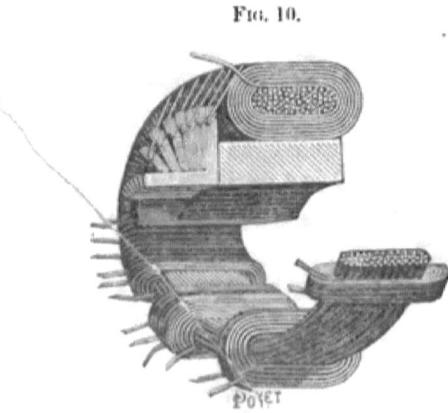

Fig. 10.

Ordinary Gramme Bobbin.

The iron wire ring is known as the Gramme ring.

The copper wire wound round the ring is called the induced wire.

The cylinder of blades is called the collector.

FIG. 11.

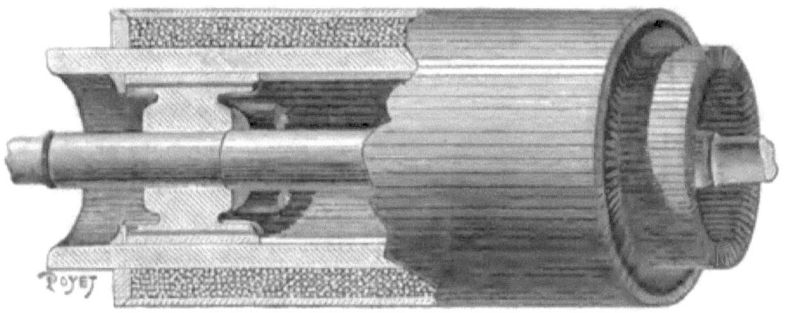

Gramme Bobbin of large capacity.

FIG. 12.

Gramme Machine for electroplating, **No. 1.**

The whole arrangement of the ring, the induced wires, the collector, the intermediary mandril, and of the washers used for the mounting on the spindle is generally called the bobbin.

This bobbin, the essential part of all Gramme machines, is illustrated in Fig. 10.

The induced wire is divided into a certain number of equal parts constituting as many elementary bobbins. Each elementary bobbin is connected at one end to one of the blades or segments of the collector, and at its other end to the following segment of the same collector. The wire is insulated by means of one or more strands of cotton passed through a bath of bitumen or of gum lac. The segments of the collector are also isolated from each other so that the ring is wound with an endless wire of great length.

Fig. 13.

Cut-out.

In order to produce machines of great capacity and low electromotive force, M. Gramme has altered the construction of his bobbin in the following manner (Fig. 11):— a series of long copper bars, isolated from each other and constituting a complete cylinder, are placed on two copper washers keyed on the spindle and having a certain number of notches. The two ends of this cylinder act as collectors. The iron wire constituting the ring is wound on this cylinder, and a second series of copper bars placed externally to the ring complete the bobbin. The internal and external bars are connected by means of radiating pieces so as to form an endless conductor as in the case of bobbins made with wire wound round the ring. The Gramme machines for electroplating are of extremely varied forms. The No. 1 type illustrated in Fig. 12 is most generally used. All these machines, without a single exception, have their electro-magnets in circuit, as they can be driven at variable speeds, and when excited in shunt or by double wire destructive sparks accompany the variations of speed.

In order to avoid the reversals of current resulting from the polarisation of the electrodes dipping in the electrolyte,

M. Gramme has devised an apparatus called the cut-out (Fig.13), which automatically cuts the circuit open when the main current becomes weakened to **too** great **an** extent by the inverse **current.** The working of this apparatus, which **is based** on the antagonistic actions of an electro-magnet and **a** spring, is too simple to require **any** description, and can be easily understood by a mere reference to the illustration.

TABLE OF GRAMME MACHINES FOR ELECTROPLATING (CURRENT TYPES).

No. of Types.	Maximum Amperes.	Corresponding Volts.	Internal Resistance.	Maximum Number of Revolutions per Minute.	Dimensions in Metres.			Diameter of the Pulley.	Width of the Pulley.	Weight in Kilogrammes.	Price in Francs.	Cost of Packing.
					Length, Pulley Included.	Width.	Height.					
1	300	15	0·008	1500	0·730	0·415	0·600	0·150	0·120	300	2000	40
2	150	18	0·024	1700	0·700	0·400	0·550	0·150	0·100	175	1000	20
3	60	17	0·066	2000	0·550	0·230	0·270	0·090	0·060	75	500	10

SIEMENS MACHINE.—The very rapid commercial development of the Gramme machine has given rise to a large number of similar machines, which are presented sometimes with a certain difference in the bobbins, sometimes with some new shapes of frames or of electro-magnets, and even sometimes with no other alterations than those resulting from a change in the sizes of the parts.

The most important of these machines is, without doubt, that invented by Mr. Hefner von Alteneck, **and which is** generally known **under the name** of **the Siemens machine.** Its success, which nearly equals that of the Gramme machine, is derived from its intrinsic merits, the care with which it is manufactured, and the really colossal industrial power of its godfathers, Messrs. Werner Siemens at Berlin, and the late **Sir** William Siemens, of London.

The Siemens machine, which we illustrate **(Fig.** 14), principally consists of a **central** cylindrical bobbin and two electro-magnets composed of **a** series of bars of soft iron, bent in the middle so as to as nearly as possible surround the bobbin.

The bobbin is constituted by a cylinder of iron wires, round which is wound the induced copper wire. It differs from the Gramme machine in this respect, that no portion of the wire is inside the iron core. The induced wire is divided into sections which are attached to a commutator identical with that of Gramme.

The current is collected by means of Gramme brushes pressing on the collectors, or by means of copper ribbons pressed in position by flat springs.

FIG. 14.

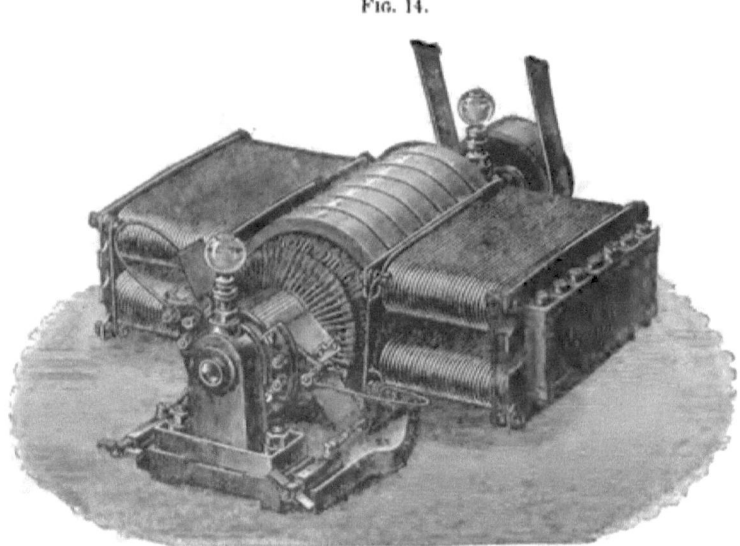

Siemens Machine.

Messrs. Siemens have manufactured a great number of types of Hefner Alteneck machines, and notwithstanding that they have found their principal market in the applications of electric lighting, they have not overlooked the problems relating to electro-metallurgy.

In the chapter treating of the refining of metals, we shall have an opportunity of describing the large machine in use at the refining works of Oker. We should have liked to give in this place a table of the machines usually employed in electroplating, but **M. Boistel**, the representative in Paris of Messrs. Siemens, informs us that the definite types are actually in course of study. We shall not be able, therefore, to publish the dimensions, weights, and prices until our second edition.

SCHUCKERT MACHINE.—**M.** Schuckert has selected as his model a Gramme machine with a flat ring and inducting polar surfaces almost entirely surrounding the induced wires, and he has established, upon that model, a series of types suitable for electric lighting and electroplating.*

FIG. 15.

Schuckert Machine.

The Schuckert machine (Fig. 15) is largely used in Germany; it is constructed with care, and the study of the types of a great capacity is particularly successful. In general, the more the maker has endeavoured to reproduce the parts and the shapes of frame adopted by the inventor, the greater real practical value of his machines.

* M. Gramme having obtained no patent in Germany for his first machine, M. Schuckert has been able to appropriate to himself the machine with flat ring without running the risk of a commercial claim on the part of the inventor.

It is to be observed that the number of volts can be increased two, three, or four fold without altering the power or the speed of the machine; this result can be obtained by winding the bobbin and the coils of the electro-magnets with a finer wire, which naturally would cause a proportional reduction in the corresponding number of amperes.

TABLE OF SCHUCKERT MACHINES FOR ELECTROPLATING.

| No. of Types. | Motive Power Required in Kilogrammetres. | Number of Revolutions per Minute. | Dimensions in Metres. | | | Diameter of the Pulley. | Width of the Pulley. | Weight in Kilogrammes. | Price in Francs. | Cost of Packing. |
			Length, including the Pulley.	Width.	Height.					
G ½	37	1000	0·73	0·27	0·35	0·065	0·040	90	575	7·50
G 1	150	800	0·90	0·38	0·43	0·100	0·065	130	800	15
G 2	225	750	1·03	0·42	0·55	0·150	0·110	200	1200	20
G 3	415	650	1·12	0·45	0·61	0·180	0·150	260	2150	30
G 4	525	600	1·33	0·50	0·68	0·210	0·175	335	2800	40
G 5	685	550	1·50	0·60	0·75	0·240	0·200	420	3250	50

The machine with a flat bobbin, having for a similar power a much larger diameter of bobbin than that of the machine with an elongated bobbin generally adopted by M. Gramme, it is evident that the number of revolutions per minute which the former can be practically driven at, is inferior to that which the latter can safely be run.

MATHER MACHINE.—Mr. Mather (of New York) preferred to borrow Siemens's bobbin to Gramme's, and, as can be seen, he has used it to great advantage. His machine (Fig. 16) has great stability, is well combined as regards the magnetism and the access to the brushes; its cost is perhaps high, but it is sold accordingly, American manufacturers buying at higher prices than those of other countries.

We have few particulars respecting the dimensions and prices of the Mather machines; we only know that machine No. 0 gives an intensity of 40 amperes, and costs 625 francs (26l.); that machine No. 1 gives 120 amperes at 1000 revolutions, and costs 1500 francs (60l.); and that machine No. 4 runs at

800 revolutions per minute, costs 10,000 francs (400*l.*), and can precipitate 10 kilogrammes of copper with 7 horse-power.

Fig. 16.

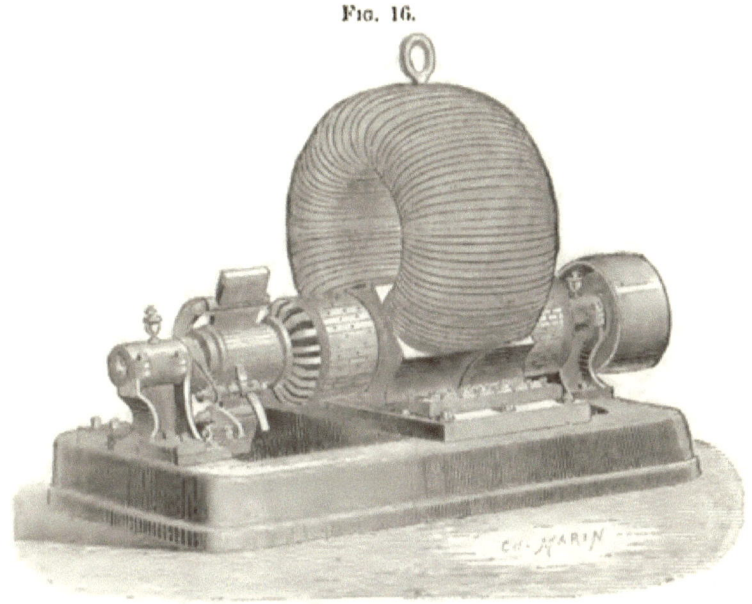

Mather Machine.

WILDE MACHINE.—The machines previously described are all continuous current machines, whereas those of Wilde, in use for more than twenty years in England, are rectified alternating current ones.

This machine is composed of two superposed machines, the top or small one exciting the electro-magnets of the bottom or large one. Both machines are provided with a Siemens longitudinal armature, that of the exciter revolving in front of permanent magnets, and that of the other in front of electro-magnets. The current of the small machine goes to excite the electro-magnets of the other, and the current of the large machine is used externally.

By means of two driving bands and a suitable engine, the two armatures are rotated, the small one at a speed of 2400, and the large one at 1500 revolutions per minute. The currents induced by the machine with permanent magnets

G

maintain the electro-magnets in a suitable state of magnetic saturation, and the currents of the principal machine are, after having been duly rectified, sent into the electroplating vats.

We shall see in another chapter that Mr. Wilde has combined a new system of dynamo, simpler and more powerful than the one which we have just described; but we wished to say a few words respecting the machine, the most perfect and the most generally used in industry before M. Gramme's conception made its triumphal appearance.

WESTON MACHINE.—The Weston machine (Fig. 17) is greatly appreciated in America, and especially in the numerous

Fig. 17.

Weston Machine.

nickeling factories of New York and Philadelphia. It is composed of a shaft armed with six radiating electro-magnets regularly distributed, and a cast-iron cylindrical casing, inside which are six stationary electro-magnets corresponding to the first mentioned.

The cylindrical casing is bolted to a bed-plate. The rotation of the spindle gives rise to alternating induced currents in the electro-magnets which are revolving with it. These currents are rectified, and go to excite the fixed electro-magnets and to produce the chemical work for which they are intended.

The Weston machine is provided with a rotary governor fixed on the bed-plate and receiving its motion from the spindle of the dynamo. The object of this governor is to automatically shunt the circuit through a resistance when the surface of the pieces in the bath becomes too small.

All the makers of electrical apparatuses also employ shunts and resistances for equilibrating the external resistance and preventing the pieces from being damaged by the action of too dense a current, but they prefer to regulate these appliances by hand instead of automatically.

The electro-magnets of the Weston machine are provided with steel plates; this arrangement prevents the change of direction of the current when the counter electromotive force becomes too great.

We are indebted for the following particulars of the Weston machines to Messrs. Hanson, Van Winkle and Co., the proprietors of the patents in the United States.

TARIFF OF WESTON MACHINES FOR ELECTROPLATING.

Diameter of the Induced armature in Metres.	Number of Revolutions per Minute.	Power required in Kilogrammetres	Dimensions in Metres.		Weight in Kilogrammetres.	Price in Francs.
			Length.	Width.		
0·15	900	5	0·35	0·25	30	750
0·20	1200	12	0·40	0·30	45	1000
0·30 ·	800	40	0·55	0·40	100	1750
0·35	500	150	0·65	0·52	230	2500
0·40	300	300	0·80	0·70	385	4000

The amount of horse-power can easily be translated into watts if we want to compare the price of these machines with those of other systems. The 40 centimetre machine, for instance, can be considered as having an efficiency of 70 per 100; with 300 kilogrammetres it therefore produces 2100 watts; its price being 4000 francs, the watt costs about 2 francs.

A Gramme machine of a value of 2000 francs gives practically 300 amperes and 15 volts, that is to say, 4500 watts, which puts the cost of the watt at 45 centimes.

A Schuckert machine of 415 kilogrammetres costs 2150

francs, which price corresponds to 74 centimes per watt, admitting, as in the case of the Weston machine, an efficiency of 70 per cent.

This calculation establishes the fact that the price of dynamos is higher in America than in Europe.

ELMORE MACHINE.—The Elmore machine (Fig. 18) has a great resemblance to the Weston; it possesses six electro-magnets revolving in front of the poles of six stationary

FIG. 18.

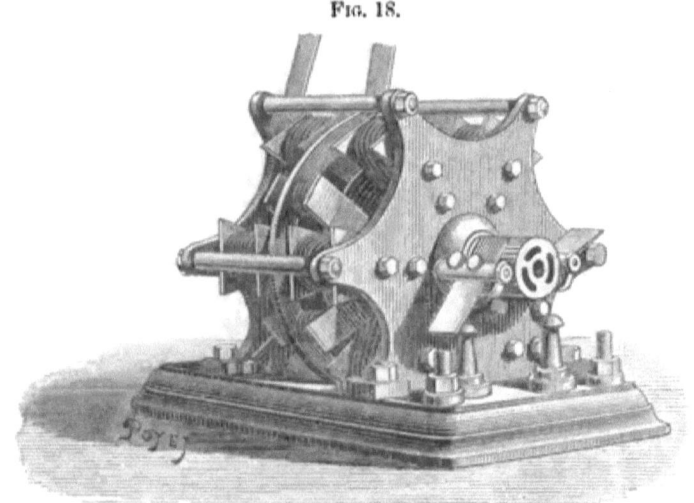

Elmore Machine.

electro-magnets and an external commutator. Only, in the Elmore machine, the two series of electro-magnets are parallel instead of being normal to each other, and each electro-magnet is composed of two coils instead of one; this is enough to constitute an original type of machine in the domain of electricity.

The Elmore machine is, besides, possessed of a particularity which is worth noticing; we refer to the cooling arrangement for preventing any part of the machine becoming overheated. To effect this the spindle of the machine is made hollow, as well as the cores of all the coils, and by means of a system of injection pipes water or cold air may be made to circulate through all the parts of the revolving disc. The refrigerating fluid is introduced at one end of the spindle, under pressure, through a stuffing-box, circulates through the whole of the induced system, and is expelled at the other end of the spindle.

The result of this cooling is to obtain a greater production of electricity for a given size of dynamo-electric machine, but it is obtained at the cost of complications in the construction and the necessity of frequent repairs; that is the reason why this arrangement has not found favour amongst the large manufacturers of dynamo-electric apparatuses. The price of Elmore machines is 2520 and 3780 francs for apparatuses requiring 1 and 2½ horse-power respectively.

MISCELLANEOUS MACHINES.—There exist a large number of other systems of dynamo-electric machines, and new types are brought out every day. These machines being mostly intended for electric lighting purposes, we will not give any description or drawings of them; they however differ very little from the types actually in use, and offer no striking peculiarity of such a nature as to give them any preponderance over these types.

We will only mention for the sake of information those which were exhibited at the Paris and Vienna Electric Exhibitions, and which appeared to us well designed:—

In the French section we noticed the Lontin machines, with inducting wheels, and which had a certain success a few years ago; and the Gerard machines, ingeniously devised for laboratory experiments and for lighting;

In the American section the Edison machines, so appreciated for lighting by incandescence, the Brush machines with rectified currents, and the Maxim machines with Gramme bobbins;

In the English section, the machines of Ferranti, Elphinstone and Bürgin for the feeding of glow lamps, and of Crompton arc lamp;

In the German section, the machines of Egger, Kremeneski, Frenzel, Gravier, Gülcher, Hauck, Jünger, Krottlinger, and Osnaghi.

SUPERIORITY OF DYNAMO-ELECTRIC MACHINES OVER BATTERIES.—We copy the following from the text of a lecture delivered before the International Congress of Electricians in 1881 by M. Henri Bouilhet, the eminent chemist of the firm of Messrs. Christofle and Co.:—

"If nickel-plating has regained favour only during the past few years, it is not because better or new solutions have been found, but because the electric industry has become enriched with a source of electricity better, more constant, and infinitely cheaper than the batteries.

"We mean the Gramme machines, which have caused a real industrial revolution, and which have rendered inestimable services in electro-chemical depositing manufactures.

"We are speaking with a certain authority, for it was in Messrs. Christofle's works that M. Gramme made his first attempt at an industrial application.

"A long time before the date of M. Gramme's invention, as far back as 1854, Messrs. Christofle had made some attempts at substituting magneto-electric machines for their batteries.

"They tried the Alliance machines, but without any appreciable success, and their price, which at that time was as high as 12,000 francs for a machine doing the same amount of work as a Gramme machine of 2000 francs would, some time later, effect, was a matter for consideration.

"Later on the Wilde machine came into practice, but its defective construction and its heating, which rendered its action intermittent instead of continuous, did not justify Messrs. Christofle in completely doing away with batteries.

"In 1871, having been brought in communication with M. Gramme, and being attracted by the principle of his constant-current machine, they placed before him the following programme for the installation of their electric current supply:—

"Construct a machine depositing 600 grammes of silver per hour on a given surface, in four baths joined in multiple arc, and revolving at a speed of 300 revolutions per minute.

"This is an occasion for paying the homage due to the inventor.

"His studies were so advanced, his calculations so certain and precise, that after three months' delay he installed at our works a machine constructed on exactly the conditions of resistance indicated by us, and at the very first trial we could deposit 600 grammes of silver per hour in the four baths, with a speed of 300 revolutions per minute.

"The service rendered by his machine to the electro-chemical industry has been so considerable, the economy in the production of work has been so great, that it has allowed of certain applications which the high price of the battery and its costly maintenance had prohibited, or which could not be thought of.

"The electro-deposition of iron, tin, nickel, and electro-metallurgic refining are proofs of it.

"In order to illustrate the importance of the industrial revolution accomplished by the Gramme machine, we will give you some interesting figures of absolutely correct cost prices, for they are established on a considerable production and on averages of five years' manufacturing work.

"With the battery, the kilogramme of silver entailed an expenditure of 3·87 francs for the galvanic current.

"With the Gramme machine, taking into account the value of the motive power, the interest of the capital, and the wear and tear of the material, the cost of depositing silver is reduced to 94 centimes per kilogramme (0·94 franc).

"This economy seems insignificant when it applies to precious metal, like gold and silver, the intrinsic value of which is considerable, but it is important when a common low price metal, like copper, tin, iron and nickel, has to be deposited.

"The success of the Gramme machine has been so consider-able that the company which works it has sold not less than 500 machines to electro-metallurgic factories alone.

"We have taken the opportunity which presented itself of speaking here of the Gramme machine, because at the very moment when the electrical exhibition displays the marvels which it has created, the progress of which it has been the starting point, and the great future which there is in store for the electrical industries using it, it has seemed to us a fit occasion for calling to mind the part, however modest, which Messrs. Christofle have taken in the industrial realisation of M. Gramme's invention."

CHAPTER VII.

CAPACITY AND EFFICIENCY OF ELECTRIC GENERATORS.

Definitions—Electromotive Force of Batteries—Maximum Yield of a Cell—Electrical Efficiency of a Battery—Maximum Capacity of a Battery composed of a certain number of Cells—Electromotive Force of Dynamo-electric Machines—Maximum Capacity of Dynamo-electric Machines.

DEFINITIONS.—The capacity of an electric generator is the quantity of useful work which it produces. It is equal to the total work generated by the apparatus, less the work absorbed by its internal resistance.

The electrical efficiency is the proportion of the capacity to the total work. For example, a machine having an internal resistance of 1 ohm and producing a current of 25 amperes with a fall of potential equal to 100 volts at the terminals has a capacity of $\dfrac{25 \times 100}{10} = 250$ kilogrammetres. The external resistance consequently is $\dfrac{100}{25} = 4$ ohms; the total work generated is $\dfrac{C^2 (r + r')}{10} = \dfrac{\overline{25}^2 \times 5}{10} = 312 \cdot 5$ kilogrammetres and the electrical efficiency is $\dfrac{250}{312 \cdot 50} = 0 \cdot 8$.

The mechanical efficiency of a machine is the proportion between its capacity and the initial work expended for keeping it in motion. If we suppose that in the preceding example the work expended was 500 kilogrammetres, the mechanical efficiency becomes $\dfrac{250}{500} = 0 \cdot 50$.

ELECTROMOTIVE FORCE OF BATTERIES.—The electromotive force of batteries is not dependent on their sizes, but solely on

the algebraic sum of the quantities of heat which they absorb
and liberate internally.

In absolute measure the electromotive force of a cell is equal
to the quantity of calories liberated per one electro-chemical
equivalent of the exciting liquid, multiplied by the mechanical
equivalent of heat.

The formula $E = 0 \cdot 0434\,H\,e$ volts (see p. 24) is applicable to batteries as well as to electrolytic baths. Let us take,
for example, the Daniell cell and observe that the chemical
work of this cell consists in the formation of sulphate of zinc
and in the destruction of sulphate of copper. The first reaction
liberates $+ 53 \cdot 5$ calories and the second one $- 28 \cdot 2$ calories,
so that the available heat, electrically speaking, is $H\,e = 53 \cdot 5$
$- 28 \cdot 2 = 25 \cdot 3$ calories. The electromotive force of a Daniell
cell is therefore $E = 0 \cdot 0434 \times 25 \cdot 3 = 1 \cdot 09$ volt.

In practice, the electromotive force is essentially variable
with cells which polarise, and it is tolerably constant, during
the life of the active substances, in cells with depolarising agents.

The following is the value of the electromotive forces of the
cells which we have described in Chapter V., the figures having
been obtained after the measurement of a great number of cells.

	Volt.
Daniell cell and its derivatives	from 0·96 to 1·08
Bunsen cell and its derivatives	„ 1·90 to 2·03
Leclanché cell	„ 1·48
Lalande and Chaperon cell	„ 0·8 to 0·9
Smee cell	„ 0·47
Clamond thermo-electric battery	„ 0·036

MAXIMUM CAPACITY OF A CELL.—The internal resistance of
batteries is dependent upon the nature of the liquids in which
the electrodes are immersed ; it also increases with the distance
of the electrodes and decreases when the immersed surfaces are
increased.

The capacity varies with the external resistance.

The capacity of a cell is maximum when the external resistance is equal to the internal resistance.

This formula, which is correct when the mode of working and
the consumption of active substances are not taken into consideration, is easy of demonstration.

Calling K the capacity, E the electromotive force, r the internal resistance, R the external resistance, and C the intensity, we have :

$$K = RC^2 = \frac{R\,E^2}{(r + R)^2}.$$

E being constant, we have to find out the maximum of the equation :

$$\frac{R}{(r + R)^2} = \frac{R}{R^2 + 2Rr + r^2} = \frac{1}{R + 2r + \dfrac{r^2}{R}}.$$

The maximum will occur when $R + 2r + \dfrac{r^2}{R}$ will be minimum ; r being constant, the problem amounts to finding the minimum of $R + \dfrac{r^2}{R}$.

That quantity is minimum when $R = r$. For if we suppose that there is, between R and r a difference a, positive or negative, the equation $R + \dfrac{r^2}{R}$ will become $r + a + \dfrac{r^2}{a + r}$, a quantity which will always be greater than $2r$, as if we subtract $r + a$ from these two equations, multiplying the results by $r + a$ we obtain respectively r^2 and $r^2 - a^2$. The second expression $2r$ is therefore inferior to the first one, and the minimum of $R + 2r + \dfrac{r^2}{R}$ which, as we have seen, corresponds to the maximum capacity, will be obtained when the external resistance R is equal to the internal resistance r.

ELECTRICAL EFFICIENCY OF A BATTERY.—The electrical efficiency of a battery is greater in proportion to the external resistance.

In the case of the maximum capacity it is :

$$\frac{R\,C^2}{(r + R)\,C^2} = \frac{r}{2r} = 0\cdot50,$$

which amounts to saying that when the battery produces the greatest possible amount of external work, it consumes internally an equal amount of work. The maximum electrical efficiency occurs when the external resistance is equal to infinity; it is then equal to unity; but the capacity becomes nil, since $E^2 \times \frac{1}{\infty} = 0$.

The minimum efficiency, on the contrary, corresponds to an external resistance equal to zero, the efficiency and the capacity both becoming nil.

In practice it is always advantageous to close a battery on a comparatively great external resistance, in order to avoid the development of heat, which is due not only to the current, but also to certain chemical reactions, which, under the influence of the temperature, quickens the dissolution of the positive electrode without increasing the electrical capacity. This, however, should not be pushed to an extreme, for almost every battery consumes a certain quantity of material per hour in either open or closed circuit, and this in excess of the quantity consumed for the generation of electricity proper; if, therefore, the normal working gives only a feeble electrical capacity, the efficiency may become insignificant, and the price of watts expensive, notwithstanding the use of cells of great electromotive force working with economical materials.

MAXIMUM CAPACITY OF A BATTERY COMPOSED OF A CERTAIN NUMBER OF CELLS.—We will suppose that a battery is composed of n cells each having an electromotive force E and an internal resistance r, and that we intend arranging this number of cells in p groups joined in series, each group containing $\frac{n}{p}$ cells joined in multiple arc for acting on a circuit of a given resistance R', giving to the battery its maximum capacity.

The internal resistance of the battery will be $\frac{p^2 r}{n}$. The electromotive force of the entire battery will be $p\,E$. The intensity of the current will be:

$$C = \frac{p\,E}{\frac{p^2 r}{n} + R} = \frac{n\,p\,E}{p^2 r + n\,R}.$$

The maximum of this equation will occur when

$$R = \frac{p^2 r}{n},$$

that is to say, when the resistance of the entire battery is equal to the external resistance.*

The quantities R, n, and r being known, it is easy to determine p, and arrange the battery with a view to obtain a maximum capacity.

ELECTROMOTIVE FORCE OF DYNAMO-ELECTRIC MACHINES.— The electromotive force of dynamo-electric machines depends on the length of the induced wire, on the speed, on the strength of the magnetic field, on the internal and external resistance of the circuit.

For a given machine it is sensibly proportional to its speed of rotation, when the inductor is a permanent magnet, or if the excitation is produced by an independent current. With a machine having electro-magnets in circuit, this proportion is not quite so exact.

In order to illustrate the variations of electromotive force obtained from variations of speed, we have carried out two series of experiments on two Gramme machines, one with permanent magnets, the other with electro-magnets; the results are shown in the following tables:—

1. MACHINE WITH PERMANENT MAGNET.

Speed of the Bobbin.		Number of Volts.	Observations.
At the Circumference.	Number of Revolutions per Minute.		
Metres.			
4·90	700	1	The circuit was closed on variable resistances, so as to give the current a constant intensity of 7 amperes.
5·60	800	2·1	
6·30	900	3·1	
7·00	1000	4·2	
8·40	1200	5·4	
10·50	1500	8	
12·60	1800	10·5	
14·70	2100	13	

* The demonstration of this proposition requires some rather too elaborate calculations to find room in a work which is especially intended for manufacturers.

2. Machine with Electro-Magnet.

Speed of the Bobbin.		Number of Volts.	Observations.
At the Circumference.	Revolutions per Minute.		
Metres.			
5·35	500	1·5	The circuit was closed on
6·42	600	2·5	variable resistances, so as to
7·49	700	3·4	give the current a constant
8·56	800	4·2	intensity of 42 amperes.
9·63	900	4·9	
10·70	1000	5·5	
16·00	1500	9	
21·41	2000	13	

Maximum Capacity of Dynamo-Electric Machines.—In the selection of a dynamo-electric machine one is guided more by its practical efficiency than by its capacity; there are, however, some instances where the maximum capacity, that is to say, the quantity of watts which a machine can industrially deliver, is of a superior interest to the efficiency; for example, in the case of a factory established on powerful and regular watercourses, a machine which costs 10,000 francs and produces 30 electric horse-power with an efficiency of only 50 per cent., will be preferable to another machine of the same price, but producing only 20 electric horse-power with an efficiency of 70 per cent.

For obtaining the maximum capacity of an industrial dynamo-electric machine it is necessary:—1st, to give the machine the greatest speed at which it can run regularly, continuously, and practically with safety; 2nd, to close the circuit on an external resistance such that the intensity of the current be at the same time the greatest possible, without any dangerous heating of the inducting and induced wires.

In the Gramme machines for electroplating, the maximum speed is 1500 revolutions per minute for the type No. 1, 1750 revolutions for the type No. 2, and 2000 revolutions for the type No. 3. The most suitable external resistance for these machines corresponds to the internal resistance multiplied by 7 for the type 1, 6 for the type 2, and 5 for the type 3.

Some authors, and notably M. Marchese, engineer at

Genoa, and M. Blast, professor at Louvain, assimilating
dynamo-electric machines to batteries, have written that the
maximum capacity of a machine was attained when the external
resistance equalled the internal resistance. This is incorrect
from a really practical point of view.

In certain conditions of speed it is, indeed, possible to get
the maximum of capacity to correspond to the equality of
resistances; but that speed, in a well made machine, never
corresponds to the maximum speed which it is possible to
obtain. Consequently, the machine does not give all the watts
which it is capable of industrially giving.

In order to demonstrate this assertion, we have undertaken
a series of experiments with the type of the Gramme machine
most generally in use and known as the normal type.[*]

The normal type of Gramme machine is particularly applicable to lighting; but it has also been used for the construction
of electroplating machine type No. 2. The volts and the
amperes naturally vary with the sizes of wires wound round the
bobbins and the electro-magnets.

We first experimented upon a machine capable of giving
and maintaining, without any dangerous heating, a current of
25 amperes, and we ran it at its practical maximum speed of
1750 revolutions. The internal resistance was, when cold,
1·083, and, after four hours' working, 1·136 ohm.

The work expended by mechanical friction and the driving
band, measured with a good dynamometer, was found to be 30
kilogrammetres without any sensible variation during the time
of the experiment.

The external resistance varied from 2 to 12 ohms. We did
not attempt to go below 2 ohms, as with such a resistance a
current of 51 amperes was made to pass through wires which
could only practically convey 25. With such an exceptional
intensity we could only get a few minutes' continuous work at
a time, owing to the abnormal heat developed in the bobbin
and in the electro-magnets.

[*] On the 1st July, 1884, the inventor had delivered to the industrial world
more than 4300 machines of this type. It is the electrical apparatus most
extensively used in the world.

Table 1 gives the results of 12 experiments conducted under conditions of maximum of speed.

The machine not being able to practically produce an intensity superior to 25 amperes, has a maximum capacity of 3325 watts, or 332·5 kilogrammetres, with an efficiency of 0·75. The external resistance R for this maximum capacity is equal to its internal resistance multiplied by 4·5. Had we gradually reduced the external resistance until R = r, we should have raised the temperature of the bobbin to such a degree that the insulating substances would have been carbonised and the machine destroyed in a few seconds.

The examination of the electrical and mechanical efficiencies will show that the greatest electrical efficiency corresponds to the smallest mechanical efficiency. This fact is explained by the constancy of the work absorbed by the frictions and the driving band, which work has the greater influence upon the mechanical efficiency in proportion of the external capacity being weaker.

The maximum mechanical efficiency of 0·778 corresponds to 18 amperes and 129 volts, that is to say, to a capacity of 2342 watts, and the internal resistance is the sixth of the external resistance. It is to be observed that for a maximum capacity of 3325 watts, this mechanical efficiency is 0·764, or very near the preceding one; that consequently the best use of this machine is that in which R = 4·46r; C = 25; E = 133.

This being established, we have reduced the speed so as to be able to close the circuit on an external resistance equal to that of the machine.

The new speed was 550 revolutions per minute; the work absorbed by mechanical frictions was reduced to 10 kilogram-metres per second.

We made twelve observations, causing the external resistance to vary from 3·875 ohms to 0·079 ohms. The results obtained are shown in table No. 2. The maximum capacity happened to exactly correspond with the equality of internal and external resistances; the total capacity, which in the first series of experiments was 3325 watts, was reduced to 648 watts, or five times less than with the practical normal speed of 1750 revolutions.

GRAMME MACHINE, NORMAL TYPE OF LOW INTENSITY, CLOSED ON A GREAT EXTERNAL RESISTANCE. (TABLE I.)

Nos. of the Experiments.	Volts at the Terminals E.	Intensity in Ampères C.	Resistance in Ohms.			Work in Kilogrammetres.			Efficiency.		Relation between the Internal and External Resistances.	Observations.
			Internal r.	External r'.	Total R.	Total Expended by the Motor W.	Electrical Actual (Capacity) $W' = \frac{EC}{g}$	Absorbed by the Internal Resistance W''.	Electrical $\frac{W'}{W'+W''}$	Mechanical $\frac{W'}{W}$		
1	50	4·16	1·136	12·01	13·15	52·80	20·80	2·00	·912	·394	10·54	Speed of the machine: 1750 revolutions.
2	67·5	6·02	1·136	11·20	12·34	74·73	40·64	4·09	·908	·544	9·82	Resistance of the machine when cold = 1·083.
3	87·5	8·57	1·136	10·20	11·34	113·35	75·00	8·35	·899	·661	8·95	
4	106	11·56	1·136	9·20	10·34	163·72	122·54	16·18	·883	·726	8·07	
5	120	14·16	1·136	8·19	9·33	224·18	169·92	24·26	·875	·758	7·18	Resistance of the electro-magnets 0·634.
6	129	18·01	1·136	7·16	8·30	298·42	232·20	36·22	·865	·778	6·28	
7	134	21·82	1·136	6·14	7·28	376·94	292·39	54·55	·843	·775	5·30	Resistance of the bobbin = 0·449.
8	133	26·11	1·136	5·09	6·23	454·37	347·26	77·11	·818	·764	4·46	
9	129	31·54	1·136	4·09	5·23	549·78	406·87	112·91	·783	·740	3·60	Work absorbed by mechanical friction: 30 kilogrammetres.
10	120	39·03	1·136	3·07	4·21	670·77	468·36	172·41	·731	·698	2·70	
11	114	46·70	1·136	2·44	3·58	809·89	532·38	247·51	·683	·657	2·14	
12	102	51·00	1·136	2·00	3·14	846·00	520·20	295·80	·638	·615	1·75	

GRAMME MACHINE, NORMAL TYPE OF LOW INTENSITY, CLOSED ON A RESISTANCE EQUAL TO THE INTERNAL RESISTANCE. (TABLE No. 2.)

Nos. of the Experiments.	Volts at the Terminals E.	Intensity in Amperes C.	Resistance in Ohms.			Work in Kilogrammetres.			Efficiency.		Relation between the Internal and the External Resistances.	Observations.
			Internal r.	External r'.	Total R.	Total Expended by the Motor W.	Electrical Actual (Capacity) $W' = \frac{EC}{g}$.	Absorb'd by the Internal Resistance W^2.	Electrical $\frac{W'}{W'+W^2}$	Mechanical $\frac{W'}{W}$		
1	15·5	4	1·136	3·875	5·011	17·02	6·90	1·82	·77	·35	3·4	Work absorbed by mechanical friction = 10 kilogrammetres.
2	24·5	7·5	1·136	3·400	4·536	34·76	18·37	6·39	·74	·52	2·9	
3	28·6	12·5	1·136	2·448	3·584	63·47	35·75	17·72	·68	·56	2·1	Speed of the machine: 550 revolutions.
4	29·4	16	1·136	1·837	2·973	86·12	47·04	29·08	·61	·54	1·6	
5	28·6	19·5	1·136	1·461	2·597	108·84	55·67	43·17	·56	·51	1·2	
6	28	22	1·136	1·272	2·408	126·59	61·61	54·98	·52	·48	1·1	Temperature of the laboratory: 22° Cent.
7	27	24	1·136	1·125	2·261	140·23	64·80	65·43	·50	·46	·99	
8	23	28	1·136	0·921	1·957	163·46	64·40	89·06	·42	·40	·72	Temperature of the machine after five hours' working: 50° Cent.
9	19	30	1·136	0·633	1·769	169·24	57·00	102·24	·36	·33	·55	
10	9	35	1·136	0·257	1·393	180·64	31·50	139·16	·18	·17	·22	
11	5	37	1·136	0·135	1·271	183·56	18·50	155·06	·11	·10	·12	
12	3	38	1·136	0·079	1·215	185·43	11·40	164·03	·063	·06	·07	

GRAMME MACHINE, NORMAL TYPE, OF GREAT CAPACITY. (TABLE No. 3.)

No. of the Experiments.	Volts at the Terminals E.	Intensity in Amperes C.	Resistance in Ohms.			Work in Kilogrammetres.			Efficiency.		Relation between the Internal and the External Resistances.	Observations.
			Internal r.	External r'.	Total R.	Total Expended by the Motor W.	Electrical Actual (Capacity) $W' = \frac{EC}{g}$	Absorbed by the Internal Resistance W".	Electrical $\frac{W'}{W'+W''}$	Mechanical $\frac{W'}{W}$		
1	2·29	3·33	·024	·6875	·7115	23·78	0·76	0·02	·97	·03	28·6	Speed of the machine: 1750 revolutions.
2	3·72	8·84	·024	·4205	·4445	26·47	3·29	0·18	·95	·12	17·5	
3	7·43	21·53	·024	·345	·369	40·09	16·00	1·09	·93	·40	14·4	Resistance of the electro - magnets ·0135.
4	13·73	57·08	·024	·2405	·2645	109·13	78·37	7·76	·91	·72	10·0	
5	16·59	82·15	·024	·1995	·2235	175·47	136·29	16·18	·89	·77	8·31	Resistance of the bobbin ·0105.
6	17·44	101·68	·024	·1715	·1955	225·14	177·33	24·71	·88	·79	7·16	
7	18·01	119·55	·024	·1505	·1745	272·98	215·31	34·67	·86	·79	6·27	
8	18·30	135·05	·024	·1355	·1595	314·00	247·14	43·86	·84	·79	5·62	Work absorbed by mechanical friction: 23 kilogrammetres.
9	18·59	144·66	·024	·1285	·1525	342·12	269·07	50·05	·84	·79	5·35	
10	18·87	154·04	·024	·1225	·1465	370·51	290·80	56·71	·84	·78	5·10	
11	19·16	163·76	·024	·117	·141	401·12	313·76	64·36	·83	·78	4·88	
12	18·87	174·72	·024	·108	·132	425·90	329·72	73·18	·82	·77	4·50	
13	18·59	186·83	·024	·0995	·1235	454·02	347·34	76·68	·80	·76	4·15	
14	18·59	198·82	·024	·0935	·1175	487·44	379·62	84·82	·80	·77	3·89	

The fall of potential at the terminals of the machine, which was 133 volts, came down to 27 volts.

The maximum of mechanical efficiency was 0·56; it corresponded to a capacity of 35·75 kilogrammetres and to $R = 2·1 r$. It is evident that under this regimen, the machine was very badly utilised; but the partisans of the equality of resistances could object that should wires of a suitable diameter be wound on the skeleton a much larger capacity could be obtained without increasing the value of E. Let us admit that the electromotive force obtained in the second series of experiments is sufficient for the use for which the machine is intended, and let us try to obtain it with a practical speed as great as possible; the normal type of electroplating machine is precisely constructed with such objects in view, and it is in experimenting upon it that we have established table No. 3.

The machine could give 154 amperes without any dangerous heating. The maximum working capacity was 2908 watts, with an electrical efficiency of 0·84 and a mechanical efficiency of 0·78. The speed was, as in the first case, 1750 revolutions per minute. The external resistance was five times greater than the internal resistance. The work absorbed by friction was 23 kilogrammetres.

This last series of experiments proves that the size of the wound wire has no influence on the equality of the resistances. With any given well-studied skeleton, a winding suitable for the work to be effected can be combined by using either short thick wires or long thin ones. In any case the maximum capacity will be obtained with an external resistance much superior to the internal resistance.

It is true that if on the last machine experimented upon we had stripped the electro-magnets a little, we might have made the maximum capacity correspond to the equality of resistances; but the economy would only have resulted in a few kilogrammes of wire, and the work would have been three or four times less than that obtained from a well equilibrated machine.

Nothing is more difficult than to uproot an error, and especially when it is based on some apparently convincing comparisons or on an incomplete although not false theory. We therefore

shall strongly insist upon this fact, that the principle of equality
of resistances is not to be realised in practice, that is to say, it
is impossible to construct a machine giving its maximum of
watts, considering the weight of the materials used in its construction, when closing the circuit on an external resistance
equal to its proper internal resistance.

SECTION III.

ELECTRO-DEPOSITION.

CHAPTER VIII.

NICKEL-PLATING.

Electro-deposition—Nickel—Physical Properties—Chemical Properties—Heat
due to the Formation of Double Salts—Nickel-plating—Preparation of the
Bath—Nickel Baths—Auxiliary Baths—Preparation of Pieces for Nickel-
plating—Nature of Anodes—Modus operandi—Electromotive Force of
the Current—Treatment when taken out of the Bath—Renickeling—
Duration of the Baths—Vats—Nickel-plating of Zinc—Powell's Process—
Deposition of Nickel by Single Immersion.

ELECTRO-DEPOSITION.—Electro-deposition, or electroplating,
is the *ensemble* of means used for the reproduction or the coating
of objects by means of the electric current. In the first instance
it is especially used for the reproduction of artistic works and
typographic engravings, and in the second for silvering, gilding,
nickeling, and coppering a large number of decorative objects
of general use.

The industry of metallic coating or plating being the most
important branch of electro-deposition, we will deal with that
first, commencing with nickel, the success of which has been
steadily growing during the last few years, in Europe as well as
in the United States of America.

NICKEL.—It will be advantageous before describing the operations required in order to obtain a good coating of nickel to give
some information respecting this metal as yet very little known,
and which is destined, in a very near future, for applications as

varied as interesting, either in the nature of a coating on other metals, or in a state of purity, or in various alloys.

PHYSICAL PROPERTIES.—In a state of purity, the metal nickel is hard, ductile, not easily fusible, and capable of receiving a beautiful polish. Its colour lies between whitish silver and greyish steel. Its density varies between 8·3 and 9·2; it is equal to 8·34 when cast, to 8·50 when reduced by hydrogen, and to 8·80 when in a forged state.

Its tenacity is, compared to that of iron, as 9:7. It can be easily forged, rolled, or drawn.

Nickel is magnetic at the ordinary temperature. Its magnetic property disappears at a temperature of 350° Cent. Under the influence of low magnetising forces, its magnetism is five times stronger than that of iron; with considerable magnetising forces it is five times less.

CHEMICAL PROPERTIES.—Nickel is soluble only in sulphuric, nitric, and hydrochloric acids. It combines directly with chlorine, phosphorus, sulphur, and arsenic.

Nickel is dissolved by dilute nitric acid; in concentrated nitric acid it is rendered passive.

Cold or hot air has no action on nickel, but certain alimentary substances, such as hot lard, rapidly dissolve it; beer, mustard, tea, and other infusions considerably alter its colour; the nickeling of cooking or culinary utensils, such as saucepans, spoons, forks, &c., is therefore not to be recommended.

The chemical equivalent of nickel is $\dfrac{59}{2} = 29\cdot5$.

HEAT DUE TO THE FORMATION OF DOUBLE SALTS.—The double sulphate of nickel and ammonium developes 94·8 calories per equivalent, starting from the metal $NH_3 + H$ and from the radical SO_4.

The double chloride of nickel and ammonium developes 98·65 calories per equivalent from the metal $NH_3 + H$ and the chlorine.

NICKELING.—In the following study we will take for our guides: M. Pérille. one of the best nickel-platers of Paris, who kindly put both his factory and his formulæ at our disposal; M. Pfanhauser, of Vienna, who has published on the subject of electro-deposition some **very** interesting notes; Mr. Watt, **of New York,** who undertook **to** propagate Mr. Adams's processes; **Mr.** Urquhart, the English author the most extensively consulted in all questions of electro-chemistry; Messrs. Neumann, Schwarz and Weil, who have established a **model works** at Freiburg; Mr. Elmore, the specialist manufacturer of London, &c.; without forgetting Messrs. Gaiffe, Roseleur, and Meidinger, **whose** works we have also consulted.

We have left out no interesting prescription, **whatever** country they might originate from, as we **consider that the** best executed work is the privilege of no particular nation, and that in the matter of nickel-plating as well as **in every other** industry, excellent methods are **to** be found everywhere.

A good nickeling must be white and solid. These qualities **are** dependent **upon** the three **following** conditions :—1st, a rational preparation of the baths and the purity of the salt of **nickel used;** 2nd, **a** suitable intensity and electromotive **force of the** electric current; 3rd, a judicious treatment of the pieces to be **nickeled,** before, during, and after their immersion **in** the chemical **bath.**

PREPARATION **OF THE** BATH FOR NICKEL-PLATING.—The purity of the salt **of** nickel is the primordial **condition for the** preparation of a bath giving good **deposits. It is difficult to** prepare salts **of** nickel perfectly free **from foreign substances,** such as copper, arsenic, cobalt, &c., **but they, however, exist,** and can be **obtained, and the operator must not** be induced **to** buy cheap material; he must take his supply from well-known dealers, and on the least doubt arising about **the** whiteness or solidity of **the** deposit, have his **salts and** acids submitted **to an** analytical chemist.

The selection of the water itself **is of very** important moment, and must be effected with judgment. **Waters from rivers,** sources, **wells, are** very often defective, because they contain, **in** solution, iron, iodine, calcic sulphates, and carbonates, &c.; the **best** plan is to use distilled water only. In the case **of large** baths, where **the** expenditure due to the use of distilled water might prove considerable, rain water may be substituted. The first water should not be collected, as it

contains in suspension the dust and other impurities from the
roof; and the clean rain water should be kept in glass or
earthenware vessels, or even in pine-wood troughs (oak troughs
are not suitable).

The proportions of salts contained in a bath must be
rigorously observed, if it is proposed to carry out any observation on the advantages of any given formula. A weak bath
becomes rapidly exhausted, whereas a too concentrated bath
gives rise to irregular crystallisations and deep colorations.

Above all, the salts added to a solution in order to render
it more active, must not decompose the salts of nickel, they
must contain no foreign substances, and, as much as possible, be
neutral.

There are a great number of formulæ for the composition of
a nickel bath, a few of which are given in the following table,
together with their mode of preparation, and the name of the
persons who recommend them. Mr. Adams was the first to point
out the use of the double chloride of nickel and ammonium, or of
the double sulphate of nickel and ammonium. His patents bear
date 1869, and in them he claims the two following preparations,
which are now the most extensively used in every country:—

1. *Chloride.*—Dissolve 135 grammes of pure nickel in hydrochloric acid, taking care not to have an excess of acid, and heat
gently. When the dissolution is effected, add 2·25 litres of
cold water; pour gradually some ammonia until the solution
becomes neutral to the test paper.

On the other part, dissolve 70 grammes of sal-ammoniac in
water, and mix it with the previous solution. Add cold water
so as to make 10 litres.

2. *Sulphate.*—Dissolve 135 grammes of pure nickel in a
mixture of sulphuric acid diluted in twice its weight of water,
and heat until completely dissolved. Add some water, and
neutralise the liquor with ammonia. Dissolve 70 grammes of
carbonate of ammonia, and pour the sulphuric acid with care
until the solution becomes neutral. Mix the sulphate of
ammonia with the sulphate of nickel, and add cold water to
make 10 litres. In both cases filter the liquors, or decant them
after settling.

NICKEL BATHS.

Nos.	Formulæ			Preparation.	Operators.
1	Double sulphate of nickel and ammonium .. .. Distilled water	1 10	kilogramme. litres.	Dissolve to saturation, in distilled water, the double sulphate of nickel and ammonium free from alkaline oxide of metals and alkaline earth metals, and filter after cooling.	Isaac Adams. Gaiffe. Elmore.
2	Double sulphate of nickel and ammonium Carbonate of ammonium .. .. .. Distilled water	0·400 0·300 10	kilogrammes. " litres.	Dissolve separately the two salts in a portion of the water, in a hot state. Pour slowly the solution of carbonate of ammonium into that containing the nickel, taking care not to go beyond the neutralisation (which is recognised when the litmus paper does not sensibly turn red).	**Roseleur.**
3	Sulphate, nitrate or chloride of nickel .. .. Disulphite of sodium (without smell) .. .. Distilled water	1 1 20	part. " "	Same preparation as above. Nos. 3 and 4 formulæ are given by a manufacturer who is reckoned as an authority. For that reason we publish them without alterations, wishing only to observe that the indications of sulphate, nitrate and chloride cannot be absolutely correct since the proportion of nickel is notably different in these three salts.	Pfanhauser.
4	Sulphate, nitrate or chloride of nickel .. .. Pure crystallised sal-ammoniac .. .. Distilled water	1 1 20	" " "		**Ditto.**
5	Sulphate of suboxide of nickel .. .. Chloride of ammonium .. .. .. Citric acid .. .. Distilled water	2 1 0·100 50	kilogrammes. " " litres.	Heat the whole to boiling point, then test with litmus. If the latter turns very red add some ammonia as pure and as concentrated as possible until neutralisation is reached.	Julius **Weiss.**
6	Nitrate of suboxide of nickel .. .. Solution of caustic ammonia .. .. Acid sulphite of sodium .. .. Distilled water	4 4 50 150	parts. " " "	Dissolve the disulphite in water, the nitrate of suboxide of nickel in ammonia, and mix the two solutions.	**G. Bolen.**

The various formulæ which we have given are principally based on the use of a double salt of nickel and ammonia indicated by Adams; it is true that Messrs. Becquerel and Ruolz had, long before 1869, made mention of ammoniacal baths, but the great success of the Adams process is due to the use of neutral baths.

Mr. Adams attributed the good deposition of nickel to the absence of potash or soda, whereas in reality excellent depositions can be obtained in ammoniacal baths containing salts of potash or of soda.

"The deposit of nickel," said M. Bouilhet at the Congress of Electricians, "is beautiful and durable only when it is effected in a neutral, or almost neutral, bath. As soon as the ammonia is in a state of liberty in an ammoniacal bath, the deposit becomes greyish and brittle. If its liberation is prevented the deposit remains homogeneous and glittering. The presence of soda or potash produces the same effect; but they are without any influence upon the deposit when they are in the state of neutral salts."

When nickel-plating articles of jewellery or philosophical instruments, various compositions of baths should be tried, the one giving the most satisfactory results being adopted. An alkaline nickel bath always gives more or less deep-shaded deposits, whereas a very little excess of acid develops a fine white colour. Instead of the citric acid indicated in the formula No. 5, many large establishments use hydrochloric acid chemically pure, owing to its much lower cost; the result is much the same, but we think it preferable to use citric acid, which is more easy of manipulation than hydrochloric acid.

For ironmongery and pieces of machinery, we advise following M. Pérille's indications, which have for many years insured the success of products plated by that firm, viz.:—in an enamelled vessel dissolve 80 grammes of double sulphate of nickel and ammonium for each litre of water, mixing it with the water when the latter has reached the state of ebullition. Pour the solution in the vat, filtering it at the same time, and taking care to let it thoroughly cool if the vat is lined with guttapercha.

The bath should be of from 6 to 8 degrees of Baumé's areometer. Litmus paper should scarcely redden when dipped into a bath so composed.

We could multiply the formulæ, but it would only result in perplexing the operator, and perhaps prevent him succeeding before a long series of trials. We prefer to remind him that it is neither the prescription nor the bath by themselves which insure satisfactory results, but the conscientious work of the operator, the purity of the chemical substances, the conduct of the electric current, the maintenance of the plant in good working order, and particularly the most minute cleanliness.

Contrary to almost all the other electro-plating solutions, such as those of gold, silver, copper, or brass, which contain potassic cyanide, the solution of a nickel bath should be absolutely free from that salt.

After a certain time the nickel baths become altered according to the anodes in use, turning either alkaline or acid. At the beginning we prefer the solutions having a very slightly acid reaction, this can be ascertained with the litmus paper, which should take a slightly pink-red colour.[*]

When the solution becomes too alkaline or too ammoniacal, a kind of greenish sediment is produced, which thickens the bath and gives the pieces a disagreeable yellow appearance. In the contrary case, when the solution becomes too acid, the deposit preserves its whiteness, but it adheres badly, swells, and peels off. In order to avoid these annoyances the operator must test his baths every morning before using them, and, if necessary, bring them to that state of slight acidity which we recommended.

Should the solution not reach the above-mentioned degree of concentration of 8 degrees, the deposition would take place too slowly; should it be higher than 10 degrees, the salts of nickel would crystallise, and crystals of an emerald green hue would be deposited on the walls of the vat and on the anodes.

[*] Litmus paper being continually used in the preparation of baths must always be kept in proper working condition. It must be kept in a closed vessel, separating the red from the blue, without which precautions the vapours which are constantly emitted in electro-plating establishments would rapidly render it useless.

We would also recommend the operator to pay attention to
the temperature of his baths, keeping the solutions, if not
at a uniform, at least at a sufficiently warm temperature during
the days and nights of winter. The hand being dipped in the
bath should not feel any impression of cold. If, for economical
reasons, the room is not heated at night, it is a good precaution
in the morning to heat to ebullition a portion of the solution
and pour it in the vats so as to raise the temperature of the
baths quicker than this would take place with the surrounding
air only. In a large establishment where the heating is
effected by means of steam, care must be taken not to introduce
any iron or lead pipes in the vats, as they would damage and
alter the solutions of nickel; earthenware pipes should be used.

AUXILIARY BATHS.—The baths of which we have given a
description are those in which the electrolytical work takes
place; the industry of nickel-plating requires some other baths
for the preparation of pieces; we will call them auxiliary baths,
and will describe them before enumerating the manifestations
which precede the coating proper.

Scouring Bath.—This bath, which is especially used for
rough cast iron, is composed of a solution of one quarter of a
litre of sulphuric acid in each 10 litres of hot water.

Potash Bath.—The composition of the potash bath is
almost the same in all the factories; it consists of a solution of
one kilogramme of potash in each ten litres of water. As it
is necessary to use this solution in a hot state, an iron worm
tube through which steam circulates may be conveniently
arranged at the bottom of the vat.

Cyanide Bath.—The cyanide bath contains a solution of
potassic cyanide in the proportion of 500 grammes of cyanide
to every ten litres of water. This bath is used for the removal
of any trace of oxide which might have been formed after the
scouring of copper or brass. This bath is of great importance
in nickel-plating, particularly as the solution of nickel, contrary
to ordinary solutions of cyanide of silver, exerts no dissolving
action on oxidised surfaces.

Whiting Bath.—As a substitute for the preceding bath,
which is expensive and offers some danger, M. Pérille advises

the use of a **bath** composed of two litres of sulphuric acid, **one** litre of nitric acid, **one-tenth litre of** calcined soot, and one-tenth **litre** of sodic **chloride.**

To prepare it, the nitric acid is first poured in an earthen-**ware jar,** then the sulphuric acid and the soot, and finally the salt; stir slowly, avoid breathing the fumes which **are** given off, and particularly the projections in the face **resulting** from the ebullition due to the temperature of the mixture.

This bath is only ready for use **six hours after** having been prepared, and should be revived once **a week with a little of** the four substances of which it is composed.

Hydrochloric Acid Bath.—For nickel-plating pieces **of iron** or steel, **a** bath composed **of a solution of one part of** hydro-chloric **acid for every five volumes of water can be substituted** for the potassic **cyanide** solution.

Dipping Copper Bath.—This bath **is used for** covering the pieces with a first metallic coating previous to the nickeling; it **is** composed **of**

Sulphate of copper	100 grammes.
Sulphuric acid	100 ,,
Distilled water	10 litres.

The sulphate **is** dissolved **in hot water and the acid is poured** in the mixture.

Coppering **Bath** *for* **Zinc.**—**This** bath **is** composed **as** follows:

Crystallised acetate of copper	200 grammes.
Carbonate of soda	200 ,,
Crystallised bisulphide of soda	200 ,,
Potassic cyanide	300 ,,
Distilled water	10 litres.

This solution should be energetically boiled before **being** used.

PREPARATION OF PIECES FOR NICKEL-PLATING. — Before being put **into the** electrolytic bath, the pieces should be pre-pared with the **greatest care;** upon this preliminary step often depends **the** final **success, and** we **beg to insist** upon that point, the more so that we have always **seen it** very much neglected **by** beginners.

In this, as in all other electro-plating operations, each country, and often even each manufacturer, have their own individual method.

These methods, however, only differ in certain details, and can be combined at will by the operator, according to the nature of the pieces to be plated and the degree of perfection which it is desired to obtain.

The following are indications formulated by a few experimenters.

Watt's Method.—All the pieces to be nickel-plated must be first dipped in a solution of caustic potash. As this solution grows weaker and weaker it must be continually revived or the pieces must be immersed during a longer time. Steel, iron, and brass can remain in the solution until the time of rubbing them with pumice-stone before they are put into the bath. Tin, Britannia metal, and pieces soldered by means of tin, must remain only a few minutes, as potash attacks those metals.

The pieces must be polished before being put in the bath, because it is more easy afterwards to polish the nickeled surface.

The polishing is done by means of a cylinder covered with a piece of tanned walrus skin or of ox-neck leather, &c., and very fine sand. The piece is then put through a second polisher and finished with finely pulverised lime. For steel and iron a small emery wheel or a leather wheel coated with emery and oil is used, the finishing process being the same as above.

Brass, after being polished, must lie for some time in the potash bath; it is afterwards dipped in a moderately concentrated solution of potassic cyanide, and, after a good rinsing, rubbed with a hard brush and pumice-stone, or with fine powdered brick. It is washed again, then redipped in potassic cyanide, rewashed and placed in the bath. It must be ascertained that the piece gets entirely covered after immersion; this is an essential condition of success. The pieces should not be manipulated with bare hands, but by means of an intermediary wet cloth.

Pfanhauser's Method.—Manufacturers who have no practice in electro-plating readily believe that rough metallic surfaces become polished after being **coated.** M. Pfanhauser, a great advocate of nickel depositions, objects **to** this notion so as not to deceive beginners, and says with reason that the final appearance **of the** object solely depends **on** its degree of polish before putting it into the bath. If it is dull, rough, or polished, it will come out dull, rough, or polished. It is therefore essential to prepare the pieces in the same way as it is desired to have them after immersion, so that after **a single cleaning** only they may be ready for sale.

The Viennese polishers, who are renowned for the **finish** given to their work, **use in the first** operation lime-wood disks provided with leather impregnated **with** glue **and** sprinkled with polishing powder.

For fine polishing they use disks made **of several** layers **of cloth soaked with** stearic oil **and** sprinkled with lime reduced **into** impalpable powder.

The velocity **of** the disks reaches 3000 revolutions per minute, and as these have a diameter of 30 centimetres, there results a speed of 50 **metres** per **second at** the point of contact of the brushes with the pieces.

Brass, bronze, or **copper** pieces **are first polished,** then scoured **in a boiling** lixivium **of** caustic soda. **Care must be taken to wrap the pieces with a copper wire so as to** manipulate them without touching them, **as a mere contact with** the fingers would grease **them.** They are afterwards dipped into pure boiling water, **rinsed, and** brushed **in cold** water until perfectly free from grease.

This operation can be carried out cold by washing and brushing with lime cream.

After cleansing, the pieces should be kept for **a very short time in** a tepid solution of potassic cyanide, so as **to** eliminate **the** oxidisation which has taken place on the surface of the metal. There only remains to rinse **them in a few** clean **waters** to enable them to efficiently receive the coating of nickel. **M.** Pfanhauser recommends in a special manner that the pieces should **not be allowed to** become dried by the air; as soon as they **are** prepared, they must be **put in the bath.**

When steel, cast iron, or iron objects are submitted **to the** process, it is advisable to get them lightly scoured after cleansing. This will be effected by dipping them in a bath of dilute sulphuric acid, and rinsing them in pure water before immersing them in the potassic cyanide solution which always precedes their being put in the bath.

M. Julius Weiss's Method. — M. Weiss distinguishes the case in which the metals to be nickel-plated have been previously polished from the case in which they are in a rough state. He says that in the first case the cleansing must be effected with the greatest care, as the polishing is generally obtained by means of oiled cloth, and it is important to remove every trace of greasy matters.

M. Weiss recommends, amongst other means, a lixivium of boiling soda, or a dipping into benzine, or again the lixivium of soda succeeding the dipping in benzine. When cleansing by means of benzine, an iron vessel, provided with a hermetic cover (owing to the liability to explosion of benzine), is used. It is filled up to about three-quarters, and the objects, suspended by wires, are immersed in it, and submitted to a stirring action up to the time of taking out, so as to remove any trace of grease. They are then rinsed in several waters, and separately brushed with pulverised lime with a brush and a little water over a flat wooden tub, and rinsed again.

For the cleaning of rough articles, a scouring done with care accompanied with a vigorous washing, and if necessary, with a cleansing in benzine, are sufficient. Of course the result is not so satisfactory as that obtained from a complete polishing operation, but in some cases it is found quite sufficient.

Gaiffe's Method. — Rub the pieces with a brush dipped into a hot thick decoction of Spanish whiting, water, and carbonate of soda. The cleansing is perfect when the pieces can easily be wetted by ordinary water.

For scouring copper and its alloys, it is sufficient to dip them for a few seconds in a bath composed of 10 litres of water and one kilogramme of nitric acid. Rough pieces require a bath composed of two parts water, one part nitric acid, and one part sulphuric acid.

For scouring iron, steel, and cast iron, the piece must be

immersed in the bath of diluted sulphuric acid until they become of a uniform grey hue, and then scrubbed with wet pumice
powder, which brings out **the bare** metal. When the pieces
are rough they must remain four hours in the scouring bath
and afterwards be scrubbed **with** well sifted **wet** sand. The
two operations are repeated until the **complete** disappearance
of the layer of oxide.

This process is not complicated, **and to those who know the**
excellence of M. Gaiffe's **products,** it will appear **quite unneces-**
sary to look for further improved processes.

Operations effected by M. Pérille.—The objects dealt **with by**
M. Pérille are mostly **of** polished steel, ordinary polished **cast**
iron, and polished malleable iron.

His workshop **is arranged** as it is illustrated **in** Fig. **19,**
page **114.** The manipulations before **the pieces are** put in **the**
bath are as follow:

1. Brush the pieces with petroleum, **turpentine, or benzine,**
then wipe in the sawdust of the left-hand **bins 1 and 2 ;**

2. Suspend the pieces separately and **dip them in a** bath **of**
boiling potash; after about 15 minutes' immersion, **rinse** them
in tub No. 3 **;**

3. Brush **lightly all the pieces on** all their faces in liquid
quicklime, then **rinse** well in **tubs Nos. 3 and 2 ;**

4. Scrub **the pieces** lightly **with some** almost liquid pumice
powder, then **rinse in** tubs 3, 2, **and 1, shake** and drain **dry,**
always holding the pieces by their hooks **;**

5. Dip quickly in the whiting bath so **that the pieces do not**
remain in this bath more than one or two **seconds ;** then **rinse**
in tubs 2 and 1 and shake well.

M. Pérille prepares the pieces **of** copper in **the same**
manner as those of cast iron or **steel,** with **the** following
variations:

1. **Leave** the pieces only a few minutes in the potash **;**

2. Do **not** rub with pumice powder the pieces which would
become unpolished or scratched by the powder or the brush **;**

3. **When** the pieces happen to be only partially polished,
dip them quickly in pure nitric acid and then dip them in the
whiting bath.

FIG. 13.

The dull scouring obtained by the acid is rendered bright by the whiting bath.

For rough, unpolished castings, M. Pérille recommends a few particular operations which gave him some very satisfactory results. They are as follow:

1. Dip the pieces in an ordinary scouring bath so as to whiten the cast iron, and rinse in water;

2. Scrub the pieces by means of an iron wire brush rapidly revolving (about 1200 revolutions per minute), using liquid pumice powder so as to thoroughly clean them. Rinse as indicated after each manipulation;

3. Dip the pieces in a copper bath so as to immediately protect them from atmospheric influence, then dry them by passing them through hot water and hot sawdust.

This latter operation is not absolutely necessary; we however recommend it because it is cheap, rapidly done, and prevents the oxidation of the pieces which have just been scoured and cleansed.

Elmore's Method.—Mr. Elmore insists upon a preliminary polishing, as perfect as possible, and recommends the use of drums lined with buffalo-leather soaked in Sheffield chalk, oil and tripoli, for the first operation. For removing the file marks or the unevenness, he recommends the use of Trent sand or of glassmaker's sand. After the surface has been smoothed by means of revolving drums thus covered with sand, the pieces are brushed and put through finely pulverised quicklime. Copper, brass, and Britannia metal pieces are finished by means of a circular brush made of a large number of calico disks kept together by leather disks and screws. Steel pieces are smoothed at first by means of an emery wheel, and afterwards by means of leather disks sprinkled with sand.

Before putting in the bath, Mr. Elmore's plan is to first dip brass and copper pieces in a hot bath of potash, rinse them, put them in the cyanide bath, rinse them again, and put them in the cleaning tub.

The cleaning must be effected with the greatest care and as rapidly as possible; it is important to frequently dip in pumice powder the hand which holds the piece. Precaution must be taken not to let a single point be untouched, so as to prevent any interruption in the metallic coating.

Thus cleaned, the piece must be rinsed in order to remove the pumice stone, dipped in the cyanide bath, rinsed again, and at last taken to the nickel bath.

Mr. Elmore recommends that pieces of Britannia metal or tin should not remain too long in the potash bath, as this bath is a powerful solvent of tin; nor to suspend pieces of Britannia metal or brass from the same bar in the potash bath, as the alkaline solution might become charged with tin and deposit some on the brass.

Cast iron pieces must be rinsed in cold water after coming out of the potash bath, then put in the dilute sulphuric acid scouring bath for from 20 to 60 minutes, rinsed again, cleaned with a very hard brush and sand and water, dipped in the hydrochloric acid bath and rinsed.

Although cast iron and iron can be directly nickel-plated, there is, according to Mr. Elmore, an advantage in previously coating them with a film of copper, because the nickel adheres better on copper than on iron, and because defects in the cleaning can be more easily detected on a coppered surface than on the naked surface of iron or cast iron.

The composition of the bath being known, the mode of preparation of the pieces elucidated, there remains to be found what are the most suitable conditions to be fulfilled in the electrolytic operation proper, and to successively examine the particularities which the anodes, the vats, the electric currents, &c., should present.

Nature of the Anodes.—When a manufacturer installs in his factory the necessary plant for nickel-plating pieces of his own manufacture, in which case the nickel-plating occupies a position secondary to the principal industry, the economy of motive power, of the consumption of acids, and of the labour entailed for the maintenance of the baths is not particularly sought after. His only object must be always to obtain solid and regular deposits. The economy which would result from a careful study of all the operations of nickel-plating would be very small compared to the total cost of the piece, and would, on the other hand, complicate his electric installation, so that no final advantage would result. But this is not the case with the nickel-plater, whose trade is to deposit metal on pieces manufactured and sold elsewhere, and whose means of action and experienced staff allow him to avail himself of all the progress which science or practice place in his way.

To him there is no small economy on operations effected on a large scale and frequently renewed : the motive power economised is represented by a saving in fuel, economy in labour by a reduced staff, &c. It is particularly with regard to the latter manufacturer that we will examine the question of the anodes,* advising the former manufacturer to use nickel anodes exclusively.

Soluble and insoluble anodes can be used in nickel-plating.

The soluble anodes must be of chemically pure nickel, in rolled bars preferably to cast bars, as the rolling makes uniform the porosity of the metal and regulates its dissolution in the bath.

Rolled bars are not quite so good conductors of electricity, but their surface being always large, this little discrepancy has no sensible influence on the consumption of current.

The insoluble anodes may be of platinum or carbon. When made of platinum, they are everlasting but very expensive ; made of carbon, they are cheap but readily get disaggregated, which causes disturbances in the baths, and they have to be replaced from time to time.

The foregoing are, speaking in a general manner, the constitutional qualities of nickel, platinum, and carbon anodes ; we will now examine their chemical behaviour and the influence which they exert on the electric current and the nickel bath.

It is *primâ facie* clear that the deposition of nickel will take place on all the pieces placed in the bath with either nickel or insoluble anodes ; but in the first case the anodes will, as fast as the pieces get coated with the metal, combine with the electrolytic solution, restoring to it the nickel which it liberates, whereas in the second case the bath will grow weaker in metal, since the nickel carried by the electrolysis from the

---

* See M. Pfanhauser's work on the same subject, Vienna, 1883.

baths on to the pieces will not be replaced. It will be at once seen that the electric energy whose duty, with soluble anodes, was confined to the equilibrium of the calorific action arising out of the passage of the current through the resistances of all kinds included in the circuit, will, with insoluble anodes, have besides to equilibrate the work of decomposition of the electrolyte. This supplementary expenditure of electrical energy can be estimated by calculation when the heat of formation of the electrolyte is known, or by experience in working sometimes with soluble and at other times with insoluble anodes, and keeping identical in both cases all the other conditions of the operation.

But the influence of the nature of the electrodes is particularly felt in the constancy of the bath. If nickel plates are exclusively used, it has been demonstrated in practice that the solution grows more and more alkaline. This fact can easily be verified by means of litmus paper: the bath, which is at the beginning neutral or sensibly so, will, in a short time, colour the red paper blue, and this coloration, which will grow deeper and deeper, is a proof of an excess of alkali. If it is not remedied the solution will thicken, and there will be formed a yellow insoluble precipitate of suboxide of nickel. Nothing is more simple than re-establishing the equilibrium of the bath by pouring into it a little citric acid, at the same time stirring the solution and testing it with litmus paper.

When platinum or carbon anodes are used the nickel deposited is entirely borrowed from the bath; the solution grows metallically weaker and more acid. The effect produced is therefore contrary to that which results from the use of soluble anodes; in the latter case the deposition will always be of a good white but will not adhere freely, and it will be impossible to obtain a thick coating of nickel.

The remedy for that state of things is easy to be found: it consists in introducing some carbonate of suboxide of nickel in the bath, and its neutrality reappears.[*]

* M. Pfanhauser recommends the following preparation for the carbonate of suboxide of nickel:—The carbonate of nickel, mixed with water, is pounded in a mortar, and by successive additions of water reduced to a state of syrup, which

There results from the foregoing that, in order to operate rationally without the trouble of having at every moment to neutralise the baths, it would be best to have in the same vat one portion of the anodes soluble and the other insoluble.

With a sufficiently large number of anodes it is easy to arrange them in such a manner that the acid layers of the solution mix themselves intimately and by single contact with the alkaline layers; with a few anodes, on the contrary, it will be useful from time to time to stir the solution so as to obtain a bath regular in all its parts. It is therefore sufficient to know the exact proportion of the two kinds of anodes in order to suppress all the operations of neutralisation; the best plan is to determine this proportion by direct experiments, establishing at the beginning twice as many insoluble electrodes as soluble ones, as this is very nearly the suitable proportion for the equilibrium between the two perturbations which we have mentioned.

There then only remains the choice between platinum and carbon for the insoluble anodes. Platinum is excellent, but, as we have said, its price is so high that many manufacturers hesitate to incur the expense. We think that a professional nickel-plater ought not to consider the expense, as his cost of establishment, intelligently directed, should procure him a saving of maintenance, expenditure, and time. Platinum has the advantage of always preserving its weight and keeping its value, and, finally, a series of platinum anodes does not constitute an excessive expenditure compared to the cost of a complete installation for nickel-plating.

The use of carbon anodes entails the inconvenience of a maintenance of an elaborate kind; the plates are, in time, eaten away by the current; they get weakened, cleaved, and finally crumble into dust. This dust disturbs the bath, and, depositing itself on the pieces to be plated, creates very disagreeable spots and unevenness. Some recent progress in the manufacture of carbon plates may lead to the expectation that before long plates of

---

is introduced into the acid bath. The bath is stirred continually, and dissolves the exact proportion of the syrup required for remaining neutralised. The excess of carbonate remains intact at the bottom of the bath, whence it can be extracted by decantation or siphoning, so as to be used for another neutralisation.

artificial carbon, capable of resisting the action of the current, will be sufficiently perfected to be advantageously used instead of platinum anodes. As soon as this has become an accomplished fact there will be no longer reason for using platinum anodes.

The soluble anodes of nickel naturally grow thinner and thinner; they must be replaced before they are reduced into impalpable fragments, as there would be a risk of stopping an operation in full work, which is always detrimental to the coating.

The suspension of the anodes in the bath should be effected by means of nickel wires about two millimetres diameter, in the shape of hooks; if copper wires were used, care should be taken not to completely immerse the anodes in the liquid, as any portion of the attachment which is immersed in the bath would get dissolved with the anode.

The vats should be of such a depth that the pieces to be plated should not, when completely immersed, reach, at a maximum, more than the two-thirds of their depths.

In principle, the pieces to be plated should never be immersed in the bath when the electric current is not acting, as otherwise the surface of the pieces might be slightly attacked by the bath, and an oxidation might arise which would prevent the close association of the coating with the underlying metal.

It is a generally acknowledged fact that the nickel bath with soluble anodes and iron pieces, for example, develops a secondary current which runs in a contrary direction and reduces the strength of the main current; it may even occur that the latter may be counterbalanced, destroyed and reversed by polarisation, which action interrupts or destroys the deposition of nickel. It is, therefore, important to use a sufficient electromotive force of current, and to be careful that the current always travel in the same direction. To that effect a galvanometer should always be inserted in the circuit, as it is the only means of really ascertaining what takes place in a bath. With the help of that instrument, not only will the inconvenience of a reversal of the current be avoided, but also the irregularity in the thickness of the deposit. At the time of starting, the position of the hand should be observed, and the current regulated in such a manner that the position of the said hand should **vary as** little as possible during the time the operation lasts.

The anodes and the pieces **to be** plated should not be too near each other in the bath. The distance should be regulated according **to** the profiles of the pieces; when the surfaces are nearly flat, 10 centimetres distance will be sufficient; with pieces having deep cavities and comparatively large reliefs, **a** space of 15, 20, and even 30 centimetres should **be** left between the anode and the piece to be plated.

SURFACE OF THE ANODES.—Certain operators **recommend** that the **anodes should be** of **the** same surface **as that of the pieces to be coated; some** others insist on the use of much **larger anodes.** The latter are right, especially when the installation **is a** small one, or the number and size of the pieces **to be** plated is **very** irregular. **Contrarily to** the solutions of gold and silver, the nickel solutions do not easily **dissolve** the anodes, and **if a** very large surface of **anodes is** not used the deposit **is localised** and its **colour** is dull.

If it is proposed to act **on large spherical or** cylindrical objects, it is found suitable to arrange all round the **surfaces to** be plated **a series of** anodes connected together so **as to obtain** a homogeneous deposit **on all the** parts on which **it is to be** effected. When **batteries are used** for producing the electric current, it is **necessary** that the total surface of the pieces suspended in **the bath** should be approximately equal **to the** surface of the zinc of the cells used.

Mr. Sprague gives as **a** maximum 85 square decimetres of surface of anode for every **100 litres of** solution of nickel. M. Pérille does not in practice exceed 40 square decimetres for the same capacity of bath.

Modus *Operandi.*—Before describing the mode of conducting the **electrolytic operation, we** beg to give the experimenter **one further** piece **of advice. As** soon as the pieces have been dipped into the potash bath, they must not be touched any more with the hand. **A hook** specially disposed, **and** the copper wire surrounding the pieces, allow **of their being** turned about, displaced, suspended, &c. It is only after they **have been** taken out of the bath and passed through hot sawdust that they can be touched with impunity. By ignoring this direction, it would be utterly impossible to obtain pieces without spots, for the cleanest hand always leaves a slightly greasy spot.

The pieces are suspended in the bath by means of the copper wire which surrounded them during the operation of cleaning, the anodes are also placed ready in position, and, as we have already said, the current passes at the same moment.

In order to secure good contacts,* four bars of copper are arranged on the walls of the vats parallel to each other. On two of these bars rest all the rods from which are suspended the pieces to be plated, and which are connected with the positive pole of the source of electricity (Fig. 20).

Mr. Watt recommends the increase, as much as possible, of the suspension wires when it is proposed to coat lead, tin, and even cast-iron pieces. He also advises that these wires should be cleaned after each operation, giving for this the following reasons:

After the suspension wires have been used once or twice (and especially when a strong coating has been deposited on the suspended object) the crystalline nickel, deposited on the unprepared surface of the wire, renders the latter very brittle, the deposited metal forming round the wire a tubular envelope which can easily be broken. Wires which have been used a few times for the purpose cease to be suitable, as they grow more brittle and less malleable.

In order to render them fit to be used again, they must be cleared of the nickel which covers them and annealed; the nickel is got rid of by dipping them in a solution of two parts of nitric acid, one part of sulphuric acid, and four parts of water.

Some writers advise an energetic electric current at the beginning so as to procure a better adherence of the nickel on the

---

* We consider it as essential to keep good contacts from the batteries or the machines up to the time pieces are immersed in the bath. The circuit must have no abnormal resistance; when certain parts of the conductors are not intimately connected together, it is necessary to keep them in a perfect state of cleanliness, and to carefully ascertain that there is no trace of oxide or of dust between the bars, the rods, and the suspension hooks. This precaution is too often neglected, and it is the cause of many failures.

objects to be coated ; this vigorous action must cease as soon as
the surface is completely whitened. The current is then reduced, and continued until a few bubbles of gas are seen to
ascend to the surface of the bath.

According to the thickness of the deposit which it is sought
to obtain, the pieces must remain from five minutes to a
few hours in the bath. During the operation, care must be

Fig. 20.

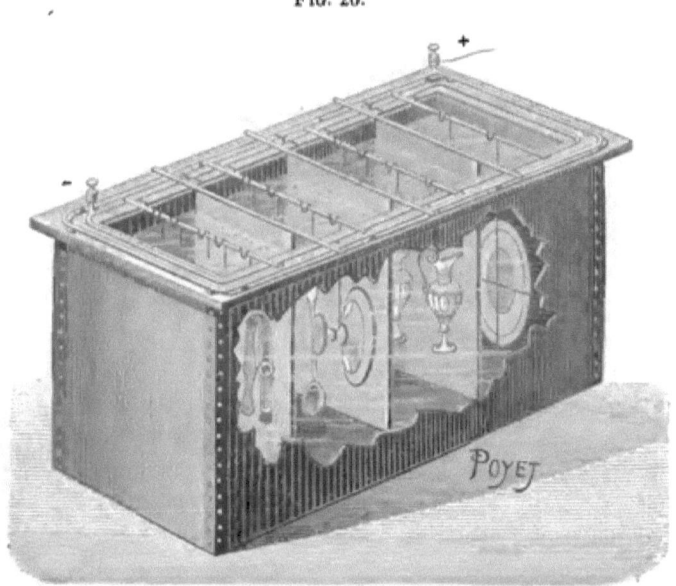

Arrangement of the Anodes and Cathodes.

taken to ascertain that the coating takes place everywhere, and
that certain parts are not black or rough. In the latter case,
the pieces should be at once taken out of the bath and the
defective parts rubbed with polishing powder, or in the lathe
with a steel wire scratch brush, and the current should be
reduced to avoid a repetition of the defect.

An excellent precaution is to frequently stir all the objects
placed in the bath. M. Pfanhauser recommends the operator
to unceasingly stir his bath so that the latter should not be still
for one instant. It is perhaps going a little far, but it is easy

to understand that the layer of liquid which surrounds the object grows weaker in metal and must frequently be replaced by a newly saturated layer. In a large factory the best arrangement consists in giving a continuous motion to the liquid by means of an automatic apparatus.

It sometimes happens that a piece to be nickel-plated is of exceptionally large dimensions compared to the sizes of the pieces generally plated; if the anodes are smaller than the said piece, and if no larger ones can be procured, or if the piece is of a very irregular profile, a sufficiently considerable distance (at least 30 centimetres) between it and the anodes should be maintained, and the operation should be very carefully watched.

If, for example, notwithstanding these precautions, the coating does not take place at the bottom of a cavity, a plate of nickel, or even of carbon, in communication with the negative pole should be introduced in the interior of that cavity.

When a large number of pieces are being plated at the same time, care must be taken that these pieces do not cover each other in the bath, and also that they are equally distributed between the anodes.

It is necessary, owing to the low conductivity of nickel solutions compared to that of gold and silver baths, to place the anodes on each side of the piece, otherwise the deposition will only be partial. For the same reason the nickel does not easily travel round angles; it is deposited freely enough on flat surfaces or on projecting parts, but with much more difficulty on concavities, internal cracks, and inside angles. This resistance to the deposition is rendered more manifest in the nickeling of cast-iron pieces, in which there are a number of holes resulting from the cores used for moulding purposes. The only means of giving a good appearance to those pieces is to first cover them with a metallic coating in a bath having a better conductivity than nickel solutions: a coating of copper for example.

Although it is often advantageous to have a thick coating of nickel, some good operators cannot obtain it without altering the deposit which cracks and peels off in large scales when it reaches a certain thickness. This phenomenon takes place when the pieces are receiving the last touch before being delivered, and the whole of the work has to be gone through again; sometimes even the nickel is seen to peel off without being touched and without a possibility of any remedy. It must be added in justice that a moderately thick coating lasts a few years, even with objects frequently in use, owing to the quite exceptional hardness of the nickel.

It is, however, not difficult to obtain a thick coating. When the bath is duly composed, the anodes well placed, and the current properly regulated, a deposition of two grammes of nickel per square decimetre can easily be effected, which corresponds to a thickness of coating equal to a fortieth of a millimetre. It is not much when the masses of the articles, some of them of daily use, are considered; it is sufficient when these coatings are compared to those of silver, so appreciated in every country. The average thickness of silver-plating executed by the best electroplaters, does not exceed three grammes per square decimetre, and as the density of silver exceeds that of nickel, it may be concluded that it is not fair to reproach nickel-plating on account of its thinness.

The truth is that, by means of successive operations of polishing and putting in bath, nothing prevents giving the nickel coating any desired thickness; the exceptional hardness of the metal alone renders useless such a series of manipulations.

From a commercial point of view, three kinds of products may be considered: 1st, articles having a bright polish; 2nd, rough articles; 3rd, cheap fancy articles.

Bright polished articles are left in the bath with a suitably regulated current until the coating turns bluish grey, slightly dull; they are then taken out and rinsed first in cold water, then in boiling water, and are finally dried in sawdust. The solidity of the coating can be ascertained by sharply and vigorously rubbing an angle of the nickel-plated object against a well planed pine board, and this until the object becomes very hot; if the nickel-plating stands the ordeal without being injured, it can be considered as excellent.

Rough unpolished metallic objects are also left in the bath until they acquire a dull bluish-grey hue. The adherence of the deposit is ascertained by scraping with a hand steel

wire brush a part out of sight, or better, on a scratch brush
lathe. Should a piece become too rough, owing to an exaggerated strength of current, it can be improved by means of a
superficial polishing and put back in the bath.

Rough and cheap articles are strung on a copper wire
and isolated from each other by means of large glass beads.
During the operation, which only lasts a few minutes, this chap-
let must be rocked to and fro without intermission. When the
objects are of very reduced sizes, as buttons, thimbles, screws,
rivets, &c., they are placed in an earthenware strainer, at the
bottom of which is a fine brass wire wound in a spiral shape;
this wire is connected with the negative pole of the electric
source. These small objects are spread about the bottom of the
strainer so as not to form too thick a layer, and are continually
stirred under the action of the current in order to alter their
situation and position. If the bath is not very conductive, the

FIG. 21.

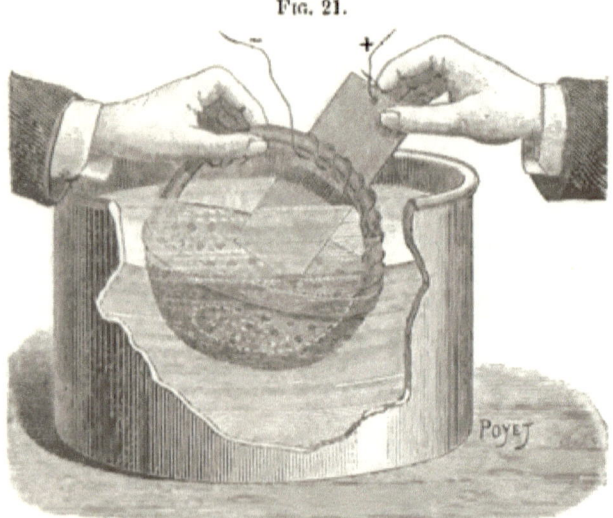

Nickel-plating of small objects.

strainer may be held in the left hand, and the anode connected
to the positive pole in the right hand (Fig. 21), taking care not
to let the anode touch the objects. This operation succeeds
very well in a hot bath, in which case it is effected quicker than

when the bath is cold, and this is particularly advantageous when the two hands are occupied.

The nickel-plating of these small objects can even be effected without the help of an external current by using a slightly acid bath and placing the objects to be coated in contact with spirals of zinc. It is then sufficient to keep the bath in a state of ebullition for a short time, stirring it with a zinc rod.

The duration of the immersion varies according to the thickness of coating which it is desired to obtain; for ordinary articles not liable to be damaged by friction, fifteen minutes will be sufficient; for ironmongery articles of current manipulation, half-an-hour is required; and for nicely finished articles an immersion of one hour seems to us to be a maximum when a Gramme machine is used. With batteries the pieces should remain two hours in the bath when an average coating is required, and five hours for a thick coating.

ELECTROMOTIVE FORCE OF THE CURRENT.—When nickel-plating was effected by means of batteries, the platers, accustomed to low-resistance baths, always joined their cells in series, and could not obtain the true colour of nickel. The deposit was of a pale yellow hue instead of having the silvery whiteness which it now possesses.

Experience shows that a good dynamo machine for nickel-plating purposes must have an electromotive force capable of varying between 1 and 8 volts. Mr. Sprague, who should always be consulted on matters of electric intensity or electromotive force, recommends beginning with 5 volts and finishing with about one. We consider it useful to reproduce the reasons which he gives for it: "The difficulty of nickel-plating does not consist in the choice of the solution but in the direction of the operation, for nickel is different from other metals in this respect, that the deposition is always accompanied by a considerable development of hydrogen, resulting in a loss of force; the object is therefore to obtain the least possible quantity of gas and the greatest possible quantity of nickel. Another consequence is that the deposition being apt to contain the gases, may become porous or scaly, in which case the coating has a tendency, as soon as it has reached a certain moderate thickness, to separate in bright pellicles. In order to prevent this escape the solution must be concentrated and the power of the battery carefully proportioned to the work to be done. For the first attack a powerful battery, such as three Bunsens in series, is required; but as soon as a general coating has been obtained economy and quality of work claim a great reduction in the electromotive force of the current. A single Smee cell, for instance, proportioned to the intensity required may be sufficient."

Mr. Sprague adds the following reflection, which we reproduce without comment, although it refers to nickel-plating by means of batteries:—" The surface of the deposited nickel, if it is good, presents a peculiar appearance; it is not bright—a bright deposit has a tendency to scaling—but its colour is of a dark yellow. After the object has been taken out of the bath and rinsed, it is brought to a bright state by means of the usual polishing processes."

TREATMENT WHEN TAKEN OUT OF THE BATH.—After being taken out of the bath the pieces must be immersed in cold water in order to rid them of any trace of sulphate, then in hot water for raising their temperature, and after in hot sawdust. If amongst the pieces there are hinges or any other apparatus subject to friction, a drop of oil should be poured on the working parts when the temperature of the pieces is comparatively high on coming out of the sawdust.

If the coating appears a little yellowish after the drying, and it is desirable to give it a beautiful white appearance, it must be polished with chalk powder or English red.

The drying in a stove is recommended for all hollow or cast objects, and particularly when they are of iron, cast iron, or steel. By placing pieces of this description in a bath of boiling oil they often become rust-proof.

The nickel-plated pieces can receive a beautiful polish, but, as we have already remarked, the metal is so hard that it is difficult to obtain the polish by the ordinary means in use. The final cleaning must be done on the lathe with brushes revolving very rapidly. An ordinary circular pig-hair brush sprinkled with soft chalk is first used, and a vigorous scrubbing given; then felt disks and finely pulverised chalk or hard red aro used, and finally a disk of wool, which gives a shining brightness.

The hollow parts are polished by means of a small rotary top, lined with cloth and sprinkled with polishing powder. The polished objects are washed freely in water, so as to rid them of the traces of wool and powder, and then dried in hot sawdust.

This polishing work is often neglected by the nickel-platers; it is, however, very cheap, and gives the pieces a shining appearance which enhances their value. The polishing of small objects nickeled in block is effected in a long and strong linen sack or in a revolving drum containing sawdust. Or they may be polished by vigorously brushing them with hot water and soap with the addition of a little ammonia.

Under certain circumstances the coating may be swollen, flaky, or scaly, notwithstanding the greatest precautions taken with a view to obtaining good results. This may happen, for instance, when the galvanometer is disturbed and a current too strong and of too great duration has been sent through the bath. If the coating cannot be perfected by following the foregoing prescriptions, it must be taken off, and the whole series of operations gone through again; this is what we call renickeling.

RENICKELING.—Renickeling is never a pleasant operation for the nickel-plater, but as it constitutes an important and necessary item in his industry it is necessary to acquaint him with the best means of proceeding.

Messrs. Watt and Elmore have treated this question with great detail, and we think it advisable to recommend their process, which is as follows:

The first operation in renickeling damaged pieces consists in stripping, by means of acids, the old coating from the piece. This is absolutely necessary, for the metal does not firmly adhere to the original coating. The following solution will generally, after a very short immersion of the pieces, take away from them all trace of nickel:

Sulphuric acid	..	..	..	..	..	4 litres.	
Nitric acid	..	..	..	..	..	500 grammes.	
Water	..	..	..	..	..	500	,,
Potassic nitrate	..	..	..	..	..	50	,,

K

The acids must be contained in a stoneware vessel which is placed under a flue provided with a good draught; the water and the potassic nitrate are added gradually. The pieces to be scoured are suspended from a copper wire and first dipped into boiling water, then quickly in the acids during half a minute. They are taken out from time to time, the operation being conducted with great caution, until the whole of the nickel coating has disappeared. The pieces thus taken out of the scouring bath must, each time before they are put back in it, be dipped in cold water; a large tub of water should be placed near the operator.

The pieces must then be thoroughly rinsed in hot water, dried and polished with great care, and put back in the nickel bath. With copper or brass pieces, if the scouring has been properly carried out, the subjacent metal should show only feeble traces of having been attacked by the acid solution. It is with a view of preventing this that sulphuric acid is introduced in the solution, and that a proportion of nitric acid only strictly sufficient for acting on the nickel is used.

The nickel can also be taken off the objects by means of the electric current by employing these objects as anodes. In this case a special bath should be used, for the solution might get spoiled by the attack of the subjacent metal, and would not be afterwards suitable for a good nickel-plating.

DURATION OF BATHS.—A nickel bath may be preserved without alteration for several years. The only precaution to be taken, when the bath is not in activity, is to provide it with a tight-fitting cover to prevent the access of dust. In summer the loss of water by evaporation is made good by the addition of a proportional quantity of distilled water at the time of using the bath.

If green crystals are deposited on the walls of the vat, it is a proof that the solution is too concentrated, and requires an addition of water. The dissolution of the crystals is facilitated by using a separate vessel containing boiling water and afterwards pouring the whole into the bath.

For getting back all the nickel from the old solutions, it is sufficient to remove the nickel electrodes and replace them by carbon or platinum electrodes, taking care to add from time to time some semi-liquid chalk and stirring frequently in order to prevent the bath becoming too acid. If the nickel of the solution is not pure enough, the objects are replaced by carbon plates, upon which the metal is very freely deposited; the said plates are afterwards dipped into nitric acid, which dissolves the metal. This solution is then boiled in a porcelain cup until complete dryness.

Mr. Urquhart recommends the following process for the extraction of nickel from old baths:—"I avail myself," says he, "of the curious property possessed by sulphate of ammonia of precipitating the double sulphates of nickel and ammonium from their solution. I therefore prepare a saturated solution of sulphate of ammonium in hot water and add it to the old solution, stirring it constantly. No result is at first to be observed, but after the lapse of a few minutes, a deposit of double sulphate begins to fall. The precipitated salt is of absolute purity, and can be directly used for making a fresh solution. This operation must be continued until the liquid has become colourless."

When the nickel solution has lost the emerald or greenish blue colour which characterises it, and has, on the contrary, become dirty, brownish or colourless, it will be well to cease using it.

If the pieces which are being plated are becoming yellow, it is because the solution has become too alkaline, and consequently requires to be neutralised by means of a weak acid, as we have already indicated.

M. Pfanhauser recommends, in order to avoid the coating becoming yellow, the following process, which seems to us to be good, and especially when a great tension is used:

As soon as the objects have been taken out of the bath, they must be immersed, without being rinsed, in a second bath specially prepared for the purpose. This supplementary bath must be acidulated by means of chemically pure hydrochloric acid, or of nitric acid. The immersion must only last a few seconds. The current, instead of being direct, will be reversed, that is to say, the anodes will be connected with the negative pole, and the pieces with the positive pole of the battery or the dynamo. The basic salts which are the cause of the yellowish tint will, by this operation, be dissolved, and the coating will become white.

VATS.—The nickeling vats must be of a capacity greater by 10 per 100 than that really required for the work to be executed. These vats are generally made of pine or pitch pine, two or three inches thick, and well joined, as shown in Fig. 22.

FIG. 22.

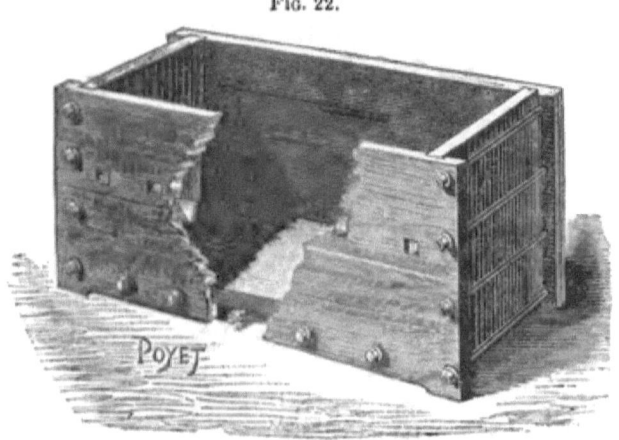

Vat for Nickel-plating.

Some vats are not lined internally; others are lined with sheets of lead soldered together by a homogeneous soldering; others are lined with guttapercha.

Mr. Sprague recommends lining the inside walls of a vat with a mixture composed of 4 parts of rosin and 1 part of guttapercha with a small quantity of boiled oil.

M. Berthoud * lines his vats with a mixture composed of 150 parts of Burgundy pitch, 25 parts of guttapercha, and 75 parts of pumice powder.

For small baths, stoneware, glass or china vessels may be used; they are more expensive and fragile, but are watertight, and easily kept clean.

Enamelled cast-iron or iron vats are also generally used in

* 'Formulaire de l'Électricien,' by E. Hospitalier.

some factories. They are suitable if the solutions are neutral or slightly alkaline, and if care is taken to well insulate them and not introduce them into circuit.

M. Brandely, in his handbook on electroplating, recommends the following recipe for rendering wooden vats watertight :

Dissolve some indiarubber and guttapercha in sulphide of carbon, reducing the solution until it is in a slightly semi-liquid state, and spread it on the wall of the vat with a large flat brush. The angles should be coated to a thickness of nearly half an inch, so as to cover any defect in the joints which often negligently occurs. In our opinion, all these coating mixtures constitute remedies which it would be better to avoid the use of, as they are not required when the work is well executed. The best vat is that which is made of thick pine well joined and lined inside with lead, ultimately covered by a thin wooden lining, the latter being held in position by means of wooden cross-bars without any screw or bolt. For small baths only, we recommend stoneware vessels, but they must be selected with care, and without any crack.

NICKEL-PLATING OF ZINC.—Zinc being easily dissolved by the solutions used for nickel-plating, it is necessary to take certain precautions for preventing pieces or fragments of zinc getting detached in the bath ; otherwise the objects suspended opposite the anodes become streaked and black, and it is utterly impossible to improve them after. As soon as a solution contains zinc, even in small quantities, it must be thrown away and a fresh one must be prepared, as the cost of extracting the zinc would be greater than that of a new bath.

It will be understood from the foregoing how carefully the nickel-plating of zinc has to be performed. The objects must first be covered with a solid copper coating, not one single point of it remaining uncovered, as without this precaution the operation fails and the bath is lost. When thus prepared, the objects can be plated without any difficulty and without any danger to the bath. When the cost is not an obstacle, it is preferable to polish the pieces before coppering them, because polishing with steel or bloodstone closes the pores of the zinc and allows the copper to more completely cover it, preserving it from the spots which are difficult to avoid without taking that precaution.

It is also necessary to take special care of the casting of pieces of zinc which are intended to be nickel-plated; the grain must be as pure and fine as possible, any abnormal cavity and the vent-holes must be hermetically closed, the surface must be perfectly homogeneous throughout, and all the pieces should be carefully examined before being put into the bath.

The recommendation never to suspend the pieces in the bath without previously passing the current applies more particularly to zinc objects; the best plan is to attach them to the conductor and actually close the circuit by the immersion of the object itself. At the beginning the current must be energetic, then it is gradually weakened after the objects are first whitened. The best arrangement would consist in having two nickel baths; a setting one, traversed by a strong current, with a large surface of anodes, and a finishing one with a low current; the two baths should of course be as neutral as possible to avoid the zinc being attacked.

The couple formed by the nickel anode and the pieces of zinc being pretty energetic, care should be taken to prevent any reversal in the current by attentively watching the galvanometer. The use of the cut-out is almost indispensable in such delicate operations.

Laminated zinc can be nickel-plated with greater facility than cast zinc, as we shall have occasion to explain in detail when we describe the works of Messrs. Neumann, Schwartz, and Weil; we shall only say now that the copper bath generally used contains an excess of potassic cyanide (300 grammes instead of 200).

Meidinger's Process.—M. Meidinger condemns the cyanide bath as being violently poisonous, and recommends the plating of the zinc by means of amalgamation.

As amalgamation renders zinc very brittle, it is necessary to leave the zinc only for a short time in the mercurial solution. If mercury is poured on a sheet of zinc previously scoured in acid, it becomes so brittle that it cannot even be bent.

The mercury used in nickel-plating exerts but a small influence on zinc when the latter is thick; but in the contrary case its action is very marked; this is the reason why the nickel-plating of zinc is a very delicate operation. It is only by actual experiments that the operator can find out the length of time required for the immersion of the zinc in the mercurial solution. The nickel deposited on amalgamated zinc is more resisting and more beautiful than that deposited on coppered zinc. M. Meidinger has observed that the various kinds of German silver were with difficulty coated with nickel, and that a better result is obtained by previously amalgamating them.

Fig. 23.

Polishing Lathe.

Works of Messrs. Neumann, Schwartz, and Weil, at Freiburg. —We illustrate, Fig. 24, one of the most complete nickel-plating works that we know of, and will give a brief description of them.

FIG. 24.

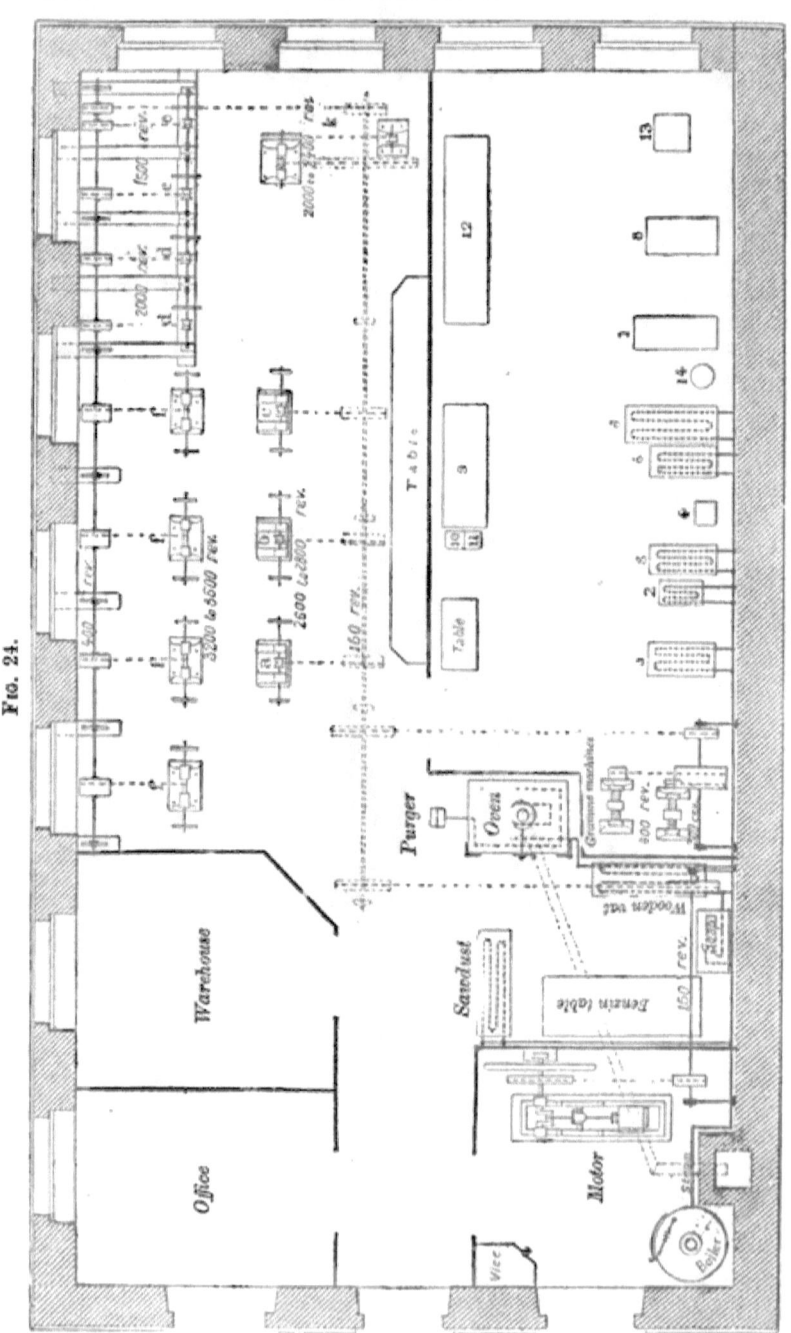

Works of Messrs. Neumann, Schwartz, and Weil.

The single room of this factory has a rectangular shape, measuring 50 feet long by 40 wide. It is divided into two parts by means of a partition which separates the bath department from the preparing and finishing departments.

The machines a, b, c, are provided at each end of their spindles with a brush used for polishing the pieces before and after the nickel-plating. These brushes are made of pieces of linen stuck over one another, and revolve at 2000 revolutions per minute. The machines d (Fig. 23), the ends of the spindles of which are terminated in points, are provided with hair brushes

FIG. 25.

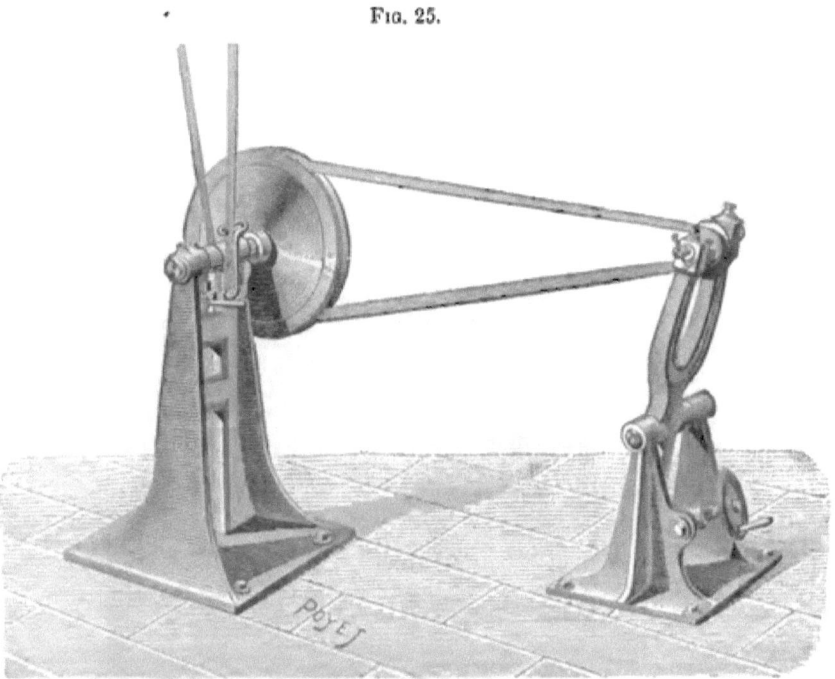

Special Polishing Machine.

revolving also at 2000 revolutions per minute, and are used for polishing pieces having slight prominences. The machines e are provided with a disk of the very hard sandstone known under the name of grisard, and on the circumference of which is stuck a leather band. This disk rotates at 1500 revolutions per minute, and is used for the polishing of plane surfaces. The ends of the spindles of the machines f are screw-shaped, so as to receive brushes, small disks, and generally all the polishers of small diameters used for the preparation of the pieces having deep recesses. The spindle rotates at 3200 revolutions per minute.

The machine h (Fig. 25) which is commonly called the special polishing machine, consists of an endless leather band, sprinkled with emery, and which can be stretched more or less by means of a screw and handle. The benzine table i is a table lined with zinc and having a raised border so as to gather the benzine which may be upset during the working.

Fig. 26.

Oven for Drying the Nickel-plated pieces.

The oven g, which is entirely built of sheet iron (Fig. 26), is used for drying the nickeled pieces and removing the benzine which remains attached to them notwithstanding their passage through the hot sawdust. **This oven is** heated by means of an iron worm **tube 2** inches in diameter.

The nickeling workshop properly speaking contains fourteen vats, as follows :

No.		Capacity.
1. Wooden box for heated sawdust	250 litres.
2. Potash bath vat in ordinary cast iron	100 ,,
3. Yellow copper bath **vat in** enamelled cast iron	**100** ,,
4. Experimental bath vat	20 ,,
5. Red copper bath vat	**480** ,,
6. Hot-water vessel	**500** ,,
7. Stoneware nickel bath **vat**	**300** ,,
8. Stoneware nickel bath vat	**200** ,,
9. Wooden cold-water vessel	**1000** ,,
10. Stoneware scouring vessel	100 ,,
11. Stoneware scouring vessel, whiting bath	100 ,,
12. Wooden nickel bath vat	2000 ,,
13. Stoneware nickel bath vat	150 ,,
14. Enamelled cast-iron experimental bath vat ..	60 ,,

The **sawdust box 1 and the bath vats 2, 3, 5 and 6 are** heated **to a temperature of 70° Centigrade by means of steam** pipes.

Amongst **the** works **executed in** this factory, we will limit ourselves to **briefly** describing the nickel-plating of sheet zinc.

Nickel-plating of Sheet Zinc.—The operations to be effected are six in number, viz. :—(1) the scouring of the rough sheets ; (2) the polishing ; (3) the cleansing ; (4) the coppering ; (5) the nickeling ; (6) the final polishing.

1. *Scouring.*—The sheets are cut to the required dimensions, **and** put in piles on a special shelf ; they are drilled in the corners and near the edges with two holes for the suspending hooks.

The workman introduces an S shape hook into each hole, and puts the sheet into the potassic bath No. 2, giving it an alternating motion during **the space of a few** seconds. He then puts it into **the water** tub No. **9,** where he dips **it** two **or** three times so as **to remove the** alkali. He afterwards puts it into the scouring bath No. 10, where it **is left one** minute, to be thereafter put into bath No. 11, and dried above vat No. 9. The sheet is then rubbed with a brush and some Spanish whiting, dipped into cold water, and put in the sawdust box No. 1 (bath No. 10 is composed of dilute sulphuric acid; bath No. 11 is composed of sulphuric and nitric acids with a small quantity of sodic chloride).

2. *Polishing.*—After being dried in sawdust, the sheet of zinc is nailed, by means of two suitable nails, to a board of the same size, and polished by machine *a*, the brush being sprinkled with saffron.

When the sheet is very bright and without scratches or spots containing traces of oxide, it is cleansed.

3. *Cleansing.*—Although very bright, the surface of the sheet of zinc contains a certain quantity of oil which must absolutely be removed. In order to obtain this result, the sheet is rubbed with some rags impregnated with benzine, then passed through sawdust, and dipped into bath No. 2, giving it a to and fro motion. It is then brushed with Spanish whiting and freely washed. These last operations must be conducted with great care, as the surface to be nickeled must not be touched with the hand.

4. *Coppering.*—The sheet, held by its hooks, is immersed in the coppering bath No. 5 during the space of twenty or thirty seconds, then taken out and successively dipped into No. 6 hot-water tub and No. 9 cold-water tub.

5. *Nickeling.*—The sheet is then passed through the nickel bath No. 13, where the deposition ought to be effected in five minutes. The nickel anode is placed very close to the sheet of zinc. The sheet is then taken out and dried.

6. *Final Polishing.*—The nickel-plated sheet is fixed to a board and submitted to the action of brushes *b* or *c*. As soon as it is sufficiently polished, it is passed through the cold sawdust, then through the benzine, and finally through the hot sawdust.

When the sheets are finished, they are piled so that the rough faces touch each other, a layer of thin paper being intercalated between the nickeled surfaces.

Powell's Process.—Mr. Powell (of Cincinnati) has a special method for obtaining nickel deposition; instead of using double salts, he uses single salts of nickel, and adds to the solution a certain quantity of benzoic acid.

He recommends the two following formulæ :

1. Sulphate of nickel   ..   ..   270 grammes.
   Citrate of nickel   200   „
   Benzoic acid   70   „
   Distilled water   ..   ..   10 litres.

2. Chloride of nickel   ..   ..   140 grammes.
   Citrate of nickel   140   „
   Acetate of nickel   140   „
   Phosphate of nickel   140   „
   Benzoic acid   70   „
   Distilled water   ..   ..   10 litres.

Benzoic acid being scarcely soluble in water, the salts of nickel should be heated in water, and when the liquid is at boiling point, add the acid, which combines with the mixture better than with water alone.

CHAPTER IX.

SILVER AND GOLD PLATING.

Importance of Silver-plating as an Industry—Composition of Silver Baths—
Concentration of Baths—Anodes—Preparation of Articles for Silver-plating
—Conduct of the Operation—Polishing of the Pieces—Observations relating
to Organic Matters—Oxidised or Antique Silver—Weakened Solutions—
Silver-plating by simple Immersion—Composition of Baths for Gold-
plating—Preparation of Articles for Gold-plating—Description of the
Process—Colouring of the Gilt Articles—Variety of Colours of the Gold
Depositions—Incrustations of Gold and Silver—Weakened Solutions—
Gilding by Single Immersion.

§ 1. SILVER-PLATING.

IMPORTANCE OF SILVER-PLATING AS AN INDUSTRY.—Silver-
plating as industrially applied was invented in 1840 by Richard
Elkington, and has long remained a kind of monopoly in the
hands of a very few firms, at the head of which can be cited
the names of Messrs. Elkington in England, and Messrs.
Christofle and Co. in France. The majority of patents have
lapsed twenty years since, and the number of manufacturers
has naturally been increasing, but not to any considerable extent,
as the important capital required by the high value of the
metal, has been a bar to many would-be manufacturers.
There are not more than ten factories in Paris where the silver-
plating business is conducted on a really industrial footing.
The small installations do not succeed, encumbered as they are
with general expenditure which prevents them advantageously
competing with more powerful and better organised firms.

The nickel-plating industry is, on the contrary, greatly
practised, not only by the professional nickel-platers, but also in
a large number of engineering workshops, where it renders
great service, without being expensive to instal.

This is no proof that nickel-plating has attained a commercial development to be compared with that of silver-plating—far from it; if the silver-plating establishments are few, they are nearly all of considerable importance, whereas the nickel-plating establishments are generally installed for operating on a small quantity of nickel at a time.

"One single silver-plating establishment in Paris," said M. Bouilhet before the Congress of Electricians, "that of Messrs. Christofle & Co., annually deposits more than 6000 kilogrammes of silver, and since the date of its foundation in 1842, it has not used less than 169,000 kilogrammes of silver deposited upon an incalculable number of objects, at a thickness suitable and sufficient for insuring to each object lasting properties appropriated to the use for which they are intended.

"The average thickness of these deposits is that which corresponds to 3 grammes per square decimetre, or 300 grammes per square metre. You will therefore see that the surface which this factory alone has covered with silver is not less than 563,000 square metres, over 56 hectares (140 acres.)

"I am only giving you here the work of one single factory; but from certain authentic information which we have been enabled to gather, we estimate at 25 tons the yearly quantity of silver used in Paris alone for silver-plating purposes.

"We have no analogous documents giving us means of ascertaining with the same degree of approximation what are the quantities of silver used in electro-deposition by other countries; but if we estimate their producing powers which are well known, and compare them with our own, it is not rash to suppose that the quantity of silver annually deposited by electrolysis in Europe and America can be computed at 125,000 kilogrammes (125 tons), which represents an intrinsic value of over 25 millions of francs (one million sterling)."

Composition of Silver Baths.—The various authors which we have consulted upon the subject do not all agree upon the question as to the best composition of a silver bath. Watt recommends a solution of argentic nitrate, or a solution formed by an electrolytic deposit of silver in potassic cyanide; Roseleur, a dilute mixture of argentic and potassic cyanide;

Japing, a solution of silver chloride dissolved in potassic cyanide; Urquhart, a mixture of silver chloride and sodic chloride; Brandely, a precipitate of argentic nitrate by potassic carbonate mixed with calcined potassium ferricyanide, &c.

Potassium ferricyanide pretty freely liberates silver, silver chloride dissolved in sodic chloride gives a chalky deposit; all solutions containing oxide, carbonate, or silver chloride give irregular results; the solution of argentic cyanide dissolved in potassic cyanide is alone exempt from inconvenience; it is therefore the only one which it is interesting to study.

This is the formula indicated by Roseleur:

Potassic cyanide of first quality	500 grammes.
Silver cyanide obtained from virgin silver ..	250 ,,
Distilled water	10 litres.

This solution is prepared in the following manner:

1. In a porcelain cup of one litre capacity put

Virgin silver in shots	250 grammes.
Pure nitric acid at 40 degrees	500 ,,

Heat the mixture on a charcoal or gas fire, placing the cup on an iron tripod in order to prevent its immediate contact with the flame.

The acid rapidly attacks and dissolves the silver, and abundantly liberates yellow fumes, the inhalation of which must be carefully avoided.

When the yellow fumes have subsided, there only remains in the cup a more or less greenish, bluish, or colourless liquid according to the proportions of copper which are contained in the silver to be obtained in the trade.

The fire is then increased in order to evaporate the excess of acid which disappears in the shape of white fumes. The matter dries and swells. A more active fire smelts it like sealing-wax. It is then taken off the fire, holding the cup with a rag, and the liquid is spread on the wall of the cup where it soon gets congealed. Silver nitrate is thus obtained.

When the whole is perfectly cold the cup is turned upside down on a piece of paper, and a slight tapping detaches the silver nitrate.

2. The silver nitrate is dissolved in a volume of distilled water equal to **ten to fifteen times its** weight, **and** hydrocyanic acid is poured in, which instantly produces **an** abundant white precipitate of argentic cyanide. It **has been** ascertained that the quantity of hydrocyanic acid poured into **the** liquid is sufficient when, on pouring a few more drops of it **into the clear** liquid which is on the top **of** the argentic cyanide, **the latter** is not further disturbed and no more precipitate is formed.

The whole is then filtered through calico ; the argentic cyanide remains on the filter, and the water, the nitric acid, and the hydrocyanic acid which might be in excess, pass through it. The precipitate remaining in the filter is then washed **in** two **or** three waters.

3. The argentic cyanide thus prepared **is poured into** the vessel which is to contain the bath, and **diluted in 10** litres of **water.** Finally the potassic **cyanide is added and is** dissolved at the same **time that it dissolves the argentic** cyanide. The double cyanide **of** potassium **and silver is thus obtained** which, as we have **said,** constitutes the **best silver-plating solution.**

CONCENTRATION OF THE BATH. — Concentrated **solutions** give a more rapid deposit than weak solutions, but require more care. The free cyanide must be nearly equal to half the weight of the dissolved silver ; with a less quantity the bath is **a bad** conductor ; with a larger **quantity,** the solution dissolves the silver of the anode and **even that already deposited on the** cathodes. Stanniferous alloys **require much more free potassic** cyanide than copper, brass, **or German silver.** When **the** anodes become coated **with a** greyish deposit, **it is a sign that** the solution is too weak of cyanide.

The solution being naturally more concentrated at the bottom than at the surface of the bath, the objects to **be** plated **receive a** greater thickness **of** deposit **at** the bottom **than** at the top ; it is therefore necessary, in **order to avoid** this inconvenience, to constantly stir the objects for the **whole** duration of the operation. This motion is, in **any** well-organised factory, mechanically obtained ; **it** is useful **not only** for silver-plating, but also in every **other** galvanic **operation.**

ANODES.—The anodes must **be of** pure silver ; they must

L

not be suspended from copper wires as those would become dissolved in the bath, but from iron wires or lead ribbons; the attachments are then joined to silvered copper conductors in connection with the source of electricity. The anodes must be entirely immersed in the bath, otherwise they would be eaten at the surface, and the inferior extremity would drop to the bottom. If the anodes are of common silver, they become red owing to the formation of cupric cyanide, which alters the purity of the deposit. The surface of the anodes must be approximately equal to the total surface of the pieces to be plated; the distance between the anodes and the pieces to be plated must be at least four inches.

PREPARATION OF THE OBJECTS.—The preparation of objects for silver-plating is, generally speaking, simpler than for nickel-plating, because the deposition of silver is effected as a rule on objects which have previously been subjected to careful workmanship, whereas the deposition of nickel is generally effected on pieces which are rough from rolling or moulding. But the operations are based on the same principle; they comprise, for copper, brass, and other similar alloys, the cleansing, pickling, scouring, and amalgamation.

1. *Cleansing.*—The pieces are placed for a certain time in a boiling solution of 10 litres of water and 1 kilogramme of potash or caustic soda, then washed first in hot water and then in pure tepid water.

2. *Pickling.*—The pickling is effected in a bath of 10 litres of water acidulated with a tenth part of sulphuric acid. A vigorous rinsing must follow.

3. *Scouring.*—The pieces are first passed through a preliminary bath composed thus:

Nitric acid at 36 degrees	10 kilogrammes.
Sodic chloride	200 grammes.
Lampblack	200 ,,

then quickly rinsed in running water and dipped in the following bath, prepared on the previous day and cooled:

Nitric acid at 36 degrees	6 kilogrammes.
Sulphuric acid at 66 degrees	200 grammes.
Sodic chloride	200 ,,

The pieces must be passed rapidly through this bath and be rinsed again in a few pure waters.

4. *Amalgamation.*—The final preparation **consists** in immersing the scoured pieces for a few seconds **in** an amalgamating liquor composed of 10 litres of **water and** 100 grammes of binoxide of mercury ; **the** pieces are stirred, and a quantity of pure sulphuric acid free of arsenic sufficient for dissolving the peroxide of **mercury** is poured into the solution. The mixture must have a limpidity equalling that of water. **After being** rinsed once more in pure water the pieces **are taken to** the silver bath.

CONDUCT OF THE OPERATION.—The deposition **is begun with a current the electromotive force of** which **does not exceed 2 to 3 volts, and the intensity 50 amperes per** square metre ; after fifteen minutes' **immersion, the** objects are taken out and inspected in **order to** ascertain **if** they are regularly coated with silver and if **no** spot or other defect is noticed ; they are brushed with tartar, rinsed, and dipped into a hot solution of potassic cyanide ; they **are rinsed again in** pure water and placed back either into the first bath or **into another silver** bath where they **are** left **until the** coating **has acquired the re-**quired thickness. In first-class firms of platers, **the deposit** of **silver on articles for the table** (table-spoons, forks) reaches from 80 **to 100 grammes** per **dozen.** The coating operation lasts three **or four hours** with the dynamo machine, and from eight to **twelve hours with a battery.**

Before taking the **objects out of the bath it** is necessary **to** interrupt the current, otherwise they **would receive a** slight yellow coloration.

POLISHING **OF THE** PIECES.—After being coated, the pieces must be dipped into a water containing some free potassic cyanide, then rinsed in boiling water and dried in sawdust (boxwood or mahogany, but not pine). The scraping of the **parts** which are **to** be bright is done by hand or in the lathe with hard hair brushes sprinkled **with** pulverised brick. The **surface is** then polished with tripoli **and red,** and afterwards burnished with special steel **or agate** tools and soaped water.

OBSERVATIONS RELATING TO ORGANIC MATTERS.—Although

a silver bath is improved by the presence of a moderate proportion of organic matter, a too sudden introduction of the
same, or its introduction in too large quantities at a time,
must be guarded against. For instance, when candlesticks,
which generally contain a mixture of rosin and pitch, are put
into the bath without having been first emptied, a large proportion of the organic substances will be dissolved by the
cyanide, and the conductivity of the bath will be reduced in a
certain proportion; the deposit will be irregular and spotty.

But when an old bath has gradually become charged with
a small quantity of organic matter, the deposits obtained
are brighter and more adhering than those obtained from a
new bath.

OXIDISED OR ANTIQUE SILVER.—The colour known as
oxidised silver is obtained as follows:

1. The silver-plated object is brushed with a camel's-hair
brush and a solution of platinum chloride in sulphuric ether,
alcohol, or cold water.

2. The following solution is then applied on it in the same
manner:

> Sulphate of copper .. .. .. 2 parts in weight.
> Potassic nitrate .. .. 1⎫
> Ammonic hydrochlorate .. .. 2⎭ dissolve in acetic acid.

3. The ammonic hydrosulphate, concentrated or dilute,
gives a more or less deep shade.

4. Sulphurous vapours give a steel-blue shade. The parts
which must not be touched should be protected by a coating.

5. Nitric acid alone produces the superficial oxidisation of
silver.

WEAKENED SOLUTIONS.—The gradual accumulation of
potassic salts which results from the action of the air upon
the free potassic cyanide comparatively rapidly alters the
silver solutions; these do not then deposit the metal with the
fine colour or the solidity with which they did it at first; it then
becomes necessary to extract the silver from the bath and
prepare a new solution.

This operation is carried out in one of the two following
manners:—The first is effected by adding acid until the metal

is completely precipitated, and melting the precipitate after drying it; it presents the serious inconvenience of emitting extremely dangerous vapours of hydrocyanic acid. The second is based on the evaporation of the solution to dryness, the fusion until complete reduction of the metal, and the elimination by lixiviation of the potassic cyanide. Mr. Sprague advocates a third process, which appears to us to be superior to the two just described; it is as follows:

"Place the solution in a large phial provided with a safety funnel tube and with an escape tube connected by means of an indiarubber tube to a large glass tube, the extremity of which dips for about 15 millimetres in a solution of argentic nitrate placed in another vessel. Add then gradually, and until a new precipitate is formed, sulphuric acid through the safety funnel tube, waiting for the effervescence to calm down, and shaking the phial. Then by means of a sand bath heat the phial and keep the solution boiling as long as a precipitate is formed in the other vessel. This precipitate is pure argentic cyanide, which if dissolved into potassic cyanide will re-constitute a new solution.

"The precipitate in the phial is also argentic cyanide, but not pure, and can be reduced by zinc and hydrochloric acid. The potassic cyanide which would have had to be replaced for precipitating the silver is thus economised."

SILVER-PLATING BY SIMPLE IMMERSION.—There exist a few solutions by means of which silver can be deposited without the help of a galvanic current; we will only describe one, as this subject is outside the scope of our book, and will select Roseleur's process as being the most efficient.

The solution recommended by Roseleur is composed of sodic bisulphite, to which is added any silver salt, but preferably argentic nitrate, until it begins to dissolve it with difficulty.

It is prepared by filling a stoneware vessel with sodic bisulphite three-fourths full, and pouring into it, stirring it all the time with a glass rod, a moderately concentrated solution of argentic nitrate in distilled water. The contact of the two liquids gives rise to the formation of large white clots of argentic sulphite, which, owing to the stirring action, the sodic bisulphite destroys, transforming them into sulphite of sodium and silver. The argentic solution is poured as long as the clots freely disappear, and the solution is ready for use when the precipitate cannot be dissolved any more. With this solution a silver deposition may be obtained from a simple whitening up to a very solid dead polish coating.

As the bath grows weaker in silver, some fresh argentic nitrate is added, and when the bisulphite has become too weak to dissolve the argentic nitrate it is sufficient to pour some into the bath in order to give the solution its original properties.

§ 2. Gold-plating.

Composition of Baths for Gold-plating.—Mr. Watt indicates five formulæ for gold baths used in the United States; they do not differ much from each other, and we will describe two of them.

1st Solution.—Dissolve in a Florentine vessel 2·33 grammes of fine gold in *aqua regia* (2 parts of hydrochloric acid and 1 part of nitric acid). Pour the solution into a porcelain cup and evaporate the acid; there remains a reddish mass of auric chloride. Dissolve this chloride, cold, in 30 grammes of distilled water. Add a concentrated solution of potassic cyanide, stirring with a glass rod, until the gold is precipitated. Decant, lixiviate the precipitate, and add some potassic cyanide for redissolving the precipitate. Evaporate to dryness in a sand bath the solution of auric cyanide, dissolve again the residuum in cold water, and filter. For use, add some boiling distilled water so as to make about 1·1 litre, and a little cyanide if the bath acts too slowly; but too much cyanide must not be used, as it would attack the anode and give the deposit a bad coloration.

2nd Solution.—Dissolve as before 2·33 grammes of fine gold and evaporate. Redissolve in 30 grammes of distilled water and precipitate the gold with ammonia, avoiding excess. Decant and lixiviate the precipitate. Dissolve the precipitate with potassic cyanide, evaporate to dryness, and dissolve again, when cold, in distilled water. Filter, and add some distilled water to make about 1·1 litre. Add a small quantity of cyanide to the solution.

M. Roselour recommends the use of two baths; one for gilding hot the small articles, the other for gilding cold the large objects.

1. Cold Process *Gilding* Solution.

Distilled water	10 litres.
Pure potassic cyanide	200 grammes.
Virgin gold	100 .,

The virgin gold, transformed into chloride, is dissolved in 2 litres and the cyanide in 8 litres of water. The two solutions are mixed, which discolours them, and boiled for 30 minutes. The effectiveness of the solution is maintained by adding equal quantities of pure potassic cyanide and auric chloride, a few grammes at a time; if the solution is too rich in gold the deposit is blackish or dark ; if it contains an excess of cyanide, the gilding is slow and the deposit grey.

2. *Hot Process Gilding Solution.*

Sodic phosphate, crystallised	600 grammes.
Sodic bisulphite	100 „
Pure potassic cyanide	10 litres.
Virgin gold transformed into chloride ..	10 litres.

Dissolve the sodic phosphate in 8 litres of hot water, let the auric chloride cool in 1 litre of water ; mix gradually the second solution with the first one : dissolve the cyanide and the bisulphite in one litre of water and mix this last solution with the two others. This bath can be used at a temperature varying from 50° to 80° Centigrade.

For dissolving gold in *aqua regia* the gold as well as the hydrochloric and nitric acids must be pure.

PREPARATION OF ARTICLES FOR GOLD - PLATING.—The articles for gilding must be prepared in the same manner as those for silvering ; that is to say, must be previously cleansed, pickled, then scoured in a preparatory bath and in the brightening bath, and lastly amalgamated, without forgetting of course the vigorous and numerous lixiviations which follow each of these operations.

Copper, brass, and generally all the cupric alloys can be directly gilded without a previous amalgamation.

Iron and steel are coppered or nickeled before being gilded. Massive silver and silver-plated pieces are gilded directly; the coppering preceding the gilding, however, consolidates the latter and renders it more durable, especially when the coating is thin. The angles rapidly become white when a layer of copper does not intervene between the two precious metals.

CONDUCT OF THE OPERATION.—Gold is possessed of very developed coating properties, so that a very fine pellicle of that metal has an appearance and gives a protection superior to that of any other metal having the same thickness. The pieces are sufficiently gilt after remaining in the bath a few minutes. The great rapidity with which the gilding is effected constitutes a great practical difficulty, as minute precautions must be taken to insure immediate success: the electromotive force must not exceed 1 volt, for the baths have a very little resistance, and the intensity per square metre of surface of cathodes must not exceed 10 amperes.

Small objects can be gilded in bulk, a stoneware sieve being used, and care being taken to constantly shake them so as to always expose new surfaces to the action of the current.

If any difficulty is experienced in obtaining the deposit in the cavities, the pieces must be thoroughly cleaned again and scratch-brushed, some cyanide must be added, and a more energetic current used until the deposit begins to be formed in the cavities. The smaller articles, like brooches, earrings, &c., must be scratch-brushed as completely as possible and stirred in the bath until the coating is uniform.

When copper and silver articles are to be gilded, they must be treated separately, beginning with the silver articles.

RAPIDITY OF THE DEPOSITION.—A bath containing 1 gramme of gold per litre, can deposit about 25 centigrammes per hour and per square decimetre.

COLORATION OF THE GILT ARTICLES.—In order to give a good appearance to gilt articles, a paste must be prepared as follows:

Alum	3 parts by weight.
Potassic nitrate	6 "
Sulphate of zinc	3 "
Sodic chloride	3 "

The objects must be dipped into this paste, or, what is better still, be brushed with it; then heated on an iron plate placed on a clear charcoal fire until they turn nearly black, and finally washed in cold water.

The following is another formula to the same effect:

Sulphate of copper	..	..	..	..	..	3	grammes.
Verdigris ..	..	..	..	..	..	7	,,
Ammonic hydrochlorate..		..	..	..	6		,,
Potassic nitrate	..	..	..	..	..	6	,,
Acetic acid	..	..	..	..	..	31	,,

Pulverise the sulphate, the hydrochlorate and the nitrate, add the verdigris and slowly pour the acetic acid, stirring at the same time. Dip the article in the preparation and heat it after on a sheet of copper until a black coloration is obtained. Let it cool, then treat it with concentrated sulphuric acid.

The article then takes a fine gold colour.

VARIETY OF COLOURS OF GOLD DEPOSIT.—By suitably mixing solutions of copper and gold, or again, solutions of silver and gold, red, pink, or green gold may be obtained.

The depositions of green gold are obtained in a bath of yellow gold in good working order, and containing from 5 to 6 grammes of gold per litre. The electric current is maintained during a few hours, a plate of pure silver being placed at the positive pole. When the colour of the metal which is deposited at the negative pole has reached the green tint which it is desired to obtain, the operation is stopped and an anode of green gold is substituted for the silver one.

Red gold is obtained in the same manner, by introducing in an ordinary gold bath a copper anode, for which is substituted a plate of alloyed gold as soon as the effect is obtained.

INCRUSTATIONS OF GOLD AND SILVER.—Messrs. Christofle and Co. have succeeded in obtaining by means of electricity some veritable damaskeen by incrusting gold and silver in bronze, iron, and steel.

The following is one of the methods employed by them:

The pattern, which later on will be in gold or silver, is executed in water-colour on the piece intended for incrustation. Water-colour easily adheres and enables the artist to at once judge of the effects which he desires to obtain. This done, the piece is protected in all its parts which are not covered by the pattern, by means of a varnish capable of resisting the action of acids and alkalies, and then placed as an anode in a weak solution of sulphuric acid. The salts of lead, of which the colours are composed, are dissolved and the metal is attacked. When the depth of the cavities so obtained is judged to be sufficient, the piece is rinsed and put in a galvanic bath of silver or gold, of low density and acting on the cold process. The deposition of the precious metal takes place and perfectly adheres in the cells which have been scoured by the action of the acid. When the cavities are full, the operation is stopped, the varnish removed, and the piece submitted to a hand polishing which reduces the excess of metal until the surfaces are level.

WEAKENED SOLUTIONS.—Gold solutions which have become altered by usage must be treated like silver solutions, by adding acid or evaporating to dryness; but the cyanide obtained and which probably contains some other metals, must be dried, mixed with a weight of litharge equal to its own weight, and melted. The residuum after lixiviation is placed in an excess of nitric acid, which dissolves the lead and leaves the pure gold.

GILDING BY SIMPLE IMMERSION.—This system is even to this day the most generally in use for imitation jewellery, cutlery, optical instruments, &c. It is effected in baths containing gold in a state of double salt of gold monoxide.

According to Roseleur the best solution is the following:

Potassic pyrophosphate	800 grammes.
Hydrocyanic acid one to eight of water	8 ,,
Crystallised auric perchloride	20 ,,
Distilled water	10 litres.

The pieces for gilding must be prepared as we have indicated for the galvanic gilding, and constantly stirred in the bath, in which they remain only a few seconds.

CHAPTER X.

COPPERING.

INTERMEDIATE COPPERING.—We have seen from the preceding that it is often useful and sometimes necessary to copper the objects which are to be nickeled, silvered, or gilt; we do not intend again to review the operations already described, but will formulate a few general recommendations upon the subject of intermediate coppering. In principle, this operation must be carried out with the greatest care in all its parts in order that the subjacent metals should entirely lose their special character; small articles or those plated in chaplets must be dipped in a hot bath so as to be rapidly coated with a film of copper. We advise taking the objects out of the bath as soon as they are coppered all over, to scrape them in order to ascertain the adherence of the precipitate, to clean the swollen parts and rub them with tartar, then to wash them and put them back into the bath.

The coppered piece must not, before receiving the coating of gold, silver, or nickel, be dried, but washed in pure water and taken to the galvanic bath after a short passage in water and the shortest possible one in air. Let us again call attention to the fact that when silvering or gilding, the coppering must be followed by a slight amalgamation, which is not required in the nickeling process.

COPPERING OF ZINC.—The industry of zinc works of art, which was a few years ago a flourishing one in France, is actually traversing a grave crisis. The foreigners who used to make their purchases in Paris, where they were certain of finding works combining taste, careful workmanship, and reasonable prices, now prefer to purchase German articles copied from our patterns, indifferently mounted, but of extremely low prices. Our manufacturers are making great exertions in order to overcome this crisis, and it is to be hoped that if they are not able to entirely stave it off they will at least succeed in diminishing its effects.

The great desideratum is to continue to produce satisfactory work and lower the prices without attempting to turn out scamped articles, otherwise our reputation would die out without any chance of being revived.

A good solution for copper baths is recommended by Mr. Watt, of New York.

It is prepared by dissolving about 230 grammes of sulphate of copper in 1 litre of hot water. Some ammonia (density ·880) is gradually added to the cooled solution, stirring well all the time until the precipitate which was previously formed is dissolved in an excess of ammonia. A concentrated solution of potassic cyanide is then poured in until the blue colour of ammoniacal sulphate of copper disappears; it is even desirable to leave an excess of cyanide; the bath is used at a temperature of 50° to 55° Centigrade.

Before being put into the bath, the pieces of zinc must be dipped into the alkaline bath and then scoured in:

Water	10 litres.
Sulphuric acid	450 grammes.

They are then passed through the sand.

M. Roseleur's formula slightly differs from the preceding one; it is as follows:

Coppering Bath.	Large Pieces.	Small Pieces.
Water	25 litres.	25 litres.
Sodic bisulphite	300 grammes.	100 grammes.
Potassic cyanide	500 ,,	700 ,,
Cupric acetate	350 ,,	450 ,,
Ammonia	200 ,,	150 ,,

Mr. Urquhart recommends as a coppering bath a solution of cupric and potassic cyanide.

"In order to easily prepare a bath," says he, "dissolve 900 grammes of potassic cyanide at 50 per cent. in 4·5 litres of water; add as much cupric cyanide as can be dissolved by the liquid, then add about 113 grammes of free potassic cyanide."

When it is desired to obtain thick deposits, it is necessary to at first apply a film in a cyanide solution, take the object out, wash it, and immediately dip it in the sulphate solution, when it can be coated to any desired thickness.

The American, French, and English formulæ which we have given are all good, and give satisfactory results. We, however, believe that that of Mr. Watt is to be preferred to a certain extent in this respect, that it is more economical and gives a more rapid deposition.

COPPERING OF CAST AND WROUGHT IRON.

OF THE IMPORTANCE OF THE INDUSTRY OF COPPERING ON CAST IRON.—The industry of artistic castings has made during the last half century considerable progress, and particularly in France, where it has succeeded in popularising the master works of statuary, and in increasing, in a large proportion, the means at the disposal of the architect for the ornamentation of our modern dwelling-houses.

Cast iron is easily melted, the mouldings are obtained with a nicety of form, a precision of outline, a neatness in the details which leave nothing to be desired. Unhappily oxidisation rapidly destroys these precious qualities, and there really exists no thoroughly effective remedy against this troublesome disease.

The best known preservative is the painting with oil of the object; but although its application is simple and economical, it must be admitted that it is not artistic, and that it requires a considerable amount of maintenance. Coppering, on the contrary, resists the action of time better, preserves the metallic appearance of the pieces which is altered by the paint, and when it is effected directly does not take away any of the artistic fineness of the cast-iron piece. Coppering has the disadvantage of being expensive and of necessitating a somewhat elaborate installation; at the same time it is injurious to the workmen's health.

Notwithstanding these inconveniences, the coppering of cast iron is beyond contest one of the most beautiful problems which has excited the inventor's imagination, and one of those of which the solutions have been the most ingenious and the most clever. The works of Messrs. Oudry, Weil, and Gauduin, amongst many others, are a proof of the labours of mankind in this particular branch of industry.

The oldest process for coppering iron is due to Baron Ruolz, and consists in decomposing the double cyanide of copper and potassium by means of a strong electric current. It gives pretty good results on very homogeneous iron and on steel on the condition that the pieces are perfectly scoured. As to ordinary iron, and especially cast iron, which always contains flaws, holes, and impurities of all kinds, the copper coating is never continuous on them, and rust soon makes its appearance at the points where the coating has not been effected. If it is sought to increase the thickness of the coating by putting the object in an acid bath of cupric sulphate, the acid, getting through the coating, attacks the subjacent metal and destroys all adherence.

This process is also an expensive one, as cyanides are of a high price, they become destroyed with time, and require a strong current to become decomposed.

OUDRY PROCESS.—After a long series of tedious and expensive experiments, M. Oudry, despairing of succeeding in depositing on cast iron a coating of copper uniting the requisite conditions of adherence, solidity, and duration, devised a mixed process consisting in coating the cast iron with a plumbagoed varnish and depositing the copper over that coating.

The varnish is composed of red lead, of resinous materials capable of thoroughly resisting the action of the air, and of plumbago which acts as a conductor to the surface. The bath may contain an acid solution of sulphate of copper without any fear of the cast iron being attacked and without the necessity of scouring it. (In general, a simple saturated solution of sulphate of copper with 10 per cent. of free acid is used.)

It is evident that a coating of copper so deposited can be possessed of no other solidity than its own, and the latter is entirely dependent on the thickness and the tenacity of the deposit. M. Oudry was accordingly led to effect depositions having one half millimetre on ordinary objects and one millimetre and more on the fine works. If to that thickness is added those of a layer of plumbago and three layers of insulating coating material, it will be readily conceived that such a system of coppering is only suitable in the case of very large objects. In the case of small objects, such as a bust for example, the nicety of the details would be irretrievably spoiled by these five layers, and it would amount to sacrificing to too great an extent the artistic worth of the object for the purpose of attaining its preservation. It is nevertheless certain that this process has really become a branch of industry, and that it is the first one which has been applied on a large scale. All the lamp-posts of the city of Paris, the beautiful fountains of the Place de la Concorde and of the Place Louvois, and a considerable number of statues and bas-reliefs have been coppered at Auteuil, in the inventor's factory.

M. Oudry's son, a few years ago, modified the original process, replacing the insulating coatings of paint and the plumbago coating by an immersion of the cast-iron objects in a thick paint composed of hot oil and copper dust in suspension in the liquid. The objects, when taken out of this bath, are dried in the oven and then rubbed with a metallic brush and some copper dust. They are finally taken to the ordinary sulphate of copper bath.

This process is rather more simple than the preceding one, but the result is practically the same. The layer of boiled oil separates the cast iron from the galvanic copper deposition; the copper dust added to the oil is only intended for rendering the surface of the pieces conductive of electricity, and the metallic deposit must always be of a certain thickness in order to be resisting. We should observe that M. Oudry senior's process was unhealthy in this respect, that it entailed the use of coating materials having red or white lead as a basis. M. Oudry junior's process also offers some dangers to the workmen's health, due to the use of dry copper dust; the latter becomes fatally introduced into the respiratory organs when the operation of scrubbing is carried on. It is said that in time of cholera this inconvenience might turn out to be an advantage; but although it is nowadays the fashion to advocate the use of copper, we cannot consider its use as healthy. In the shape of dust, particularly, it is eminently toxic, and certain precautions should be taken to avoid its dangerous effects.

In conclusion, we must add that the cost of the coppering of cast iron over an insulating layer is comparatively high, since the thickness of the deposited coating is necessarily great, and that, notwithstanding its thickness, it is liable to be deformed under the slightest shock. Many pieces get torn at the angles, and these defects have to be concealed by means of certain contrivances, the most generally used being the application, when hot, of a cement composed of rosin and copper dust on to the damaged parts. Almost all the pieces have to be mended in that fashion, and it is not one of the least inconveniences of the process.

In principle, it can be admitted that the pieces coppered by that process cost a price varying between 1 franc and 1 fr. 50 c. per kilogramme.

WEIL'S PROCESS.—M. Weil obtains a direct coppering in an alkaline liquor. This liquor consists of sulphate of copper dissolved in an excess of potassic or sodic tartrate, alkalised with caustic soda. The objects of the excess of alkali are: to keep the copper in a state of dissolution (for cupric tartar alone is insoluble in water) and to prevent the cast iron being attacked by the acid, which secures the adherence of the copper to the cast iron. The property which alkalino-organic solutions possess of dissolving ferric oxide without attacking the iron itself is availed of for the scouring of the pieces. The bath scours the iron at the same time that it coppers it.

The objects are suspended in the bath by means of zinc wires.

The current is produced by the dissolution of zinc in the alkali; the expenditure of electricity is therefore very small;

but owing precisely to the said dissolution of zinc, the bath after a certain time contains more zinc than copper, and must be frequently renewed. The copper coating, always very thin, often has a friable appearance, and becomes coloured with iridated shades due to a slight oxidisation of the copper.

M. Weil has perfected his process, and in a note recently addressed to the Academy of Sciences he thus describes the three methods which he generally uses :—

"The first method consists of dipping the pieces in the bath in contact with the zinc wires. The coppering immediately takes place, and afterwards secures the subjacent metal from the attack of acids. It requires a time varying from a few minutes to a few hours, according to the alkalinity of the bath and the purposes for which the objects are intended.

"The second method, which is used for coppering the lamp-posts of large towns, consists in placing some porous cells in the vat containing the copper alkalino-organic bath and the objects to be coated with an average thickness. These porous cells are filled with a lixivium of caustic soda, in which some zinc plates connected by means of a large zinc wire to the pieces to be coppered are immersed. The sodic lixivium can be indefinitely used, for, as soon as it is nearly saturated with oxide of zinc, it is treated by sodic sulphide, which regenerates the caustic soda, at the same time precipitating some white sulphide of zinc, which is sold at a good price. Coppering by this method requires only a short time.

"The third method applies to the coppering of various objects of small, average, or very great thicknesses, by means of the same baths and a dynamo-electric machine.

"The baths only require from time to time the addition of a given quantity of oxide of copper. When the copper is nearly exhausted, it is titrated as follows :—Put 10 cubic centimetres of the bath solution in a white glass matrass, add 30 to 40 cubic centimetres of pure hydrochloric acid ; boil and pour the greenish yellow solution on titrated stannic protochloride until complete discoloration of the latter. The volume of stannic chloride used for that purpose correctly indicates the quantity of copper contained in the bath. There only remains to add to it the quantity of hydrated oxide of copper which it requires."

M. Weil deposits all the metals, such as nickel, cobalt, antimony, tin, &c., on cast iron, iron, and other metals. He uses for this purpose alkalino-organic baths of an analogous composition to that of his coppering bath, and the operation is carried out by any of the three methods described for coppering.

Some surprise may be created by the fact that M. Weil treats the scouring of the pieces to be coppered as a matter of small importance. Surfaces so irregular and heterogeneous as those of rough forgings or castings cannot be scoured like the copper surfaces intended for silver plating; and the experimenter knows that a coating never adheres or resists well if the scouring is not well effected. However, notwithstanding this little criticism, we cannot deny that this coppering system is good, and that it is worked out with intelligence and profit.

GAUDUIN'S PROCESS.—M. Gauduin is the inventor of three processes for coppering iron and cast iron; the two-first being based on the dry method, are out of the scope of our work; the third has been thus described by the inventor himself.

"Many organic acids, and principally the polybasic organic acids, tartaric, oxalic, succinic, citric, malic, and others possessing the same properties, are more or less suitable for the production of a good deposit when they are combined with the oxides of copper and the alkaline oxides in a state of double-acid salts, as in bitartrates, bioxalates, quadroxalates, bisuccinates, bicitrates, alkaline bimalates, and with the corresponding salts of copper with an excess of acid when they do not too strongly attack iron. The operation is facilitated by a temperature of 40° to 60° C. The use of an electric current produced by any given source is necessary when the coating is to be of a great thickness. The manner in which this current is used, as well as the preliminary scouring of the iron or of the cast iron, do not require description, as these operations are not possessed of any features special to this process.

"Alloys of copper and tin, copper and zinc, copper and aluminium (bronze, brass, aluminium bronze), can also, by means of these baths, be deposited on iron, cast iron, and steel.

This result is obtained by mixing in the above-described copper bath a similar bath of tin, zinc, or aluminium, in quantities inversely proportional to the electric conductivity of the two baths, and operating with soluble anodes of bronze, brass, or aluminium bronze. The same results may be obtained by using for a certain time a copper bath with a soluble anode of bronze, brass, aluminium bronze; or better, tin, zinc, or aluminium, taking great care to regulate the tension of the current proportionally to the surface to be coated. This is an essential condition to the success of the operation, for a weak current only affects the copper, whereas a strong current preferably affects the other metals."

The Société du Val d'Osne, who are working Gauduin's processes, use a small Gramme machine, No. 2, which requires less than one horse-power for working the immense baths installed in the Boulevard Voltaire, in Paris. It was there that the first experiments on a large scale took place of the excellent chemist whose recent death we chronicle with regret.

M. Gauduin at first employed soda for the formation of his double salt. The oxalates of copper being only slightly soluble, the effect of the presence of soda was to dissolve a greater quantity of copper, rendering therefore the bath a better conductor. But, notwithstanding all his exertions, he never succeeded in constituting a bath rich enough to work in a cold state.

Heat was indispensable, and caused considerable expenditure in material and fuel.

A few years later the composition of the bath was altered. Ammonia was substituted for soda, and being an active dissolvent of copper this was the means of obtaining baths very rich in copper. A great economy was thus realised, and from that time the operation has been conducted with the greatest care and in a very regular manner.

Before being put into the bath, the piece is cleansed by means of potash, and thoroughly scoured by means of sulphuric acid. After being washed in water it is put in the copper bath.

The anodes being of copper keep the richness of the bath at an almost constant point, and it is sufficient to only add from time to time a small quantity of an ammonical dissolution of copper. The bath is gradually charged with oxalate of iron through the scouring of the pieces, and the presence of this salt, instead of interfering with the copper deposition, seems, on the contrary, to favour it and render it more regular. The same bath can therefore be indefinitely used.

When the piece has received, on all its surface, a sufficiently thick deposition of copper so as not to be influenced by the sulphuric acid, the operation is completed in a bath of sulphate of copper, simply composed of a saturated solution of about 10 per 100 of free sulphuric acid. This bath has the advantage of being more economical than the previous one, and of giving a more rapid deposition, at the same time requiring a smaller current power.

It will be understood from the foregoing that the problem of coppering cast iron is perfectly solved by the cupric ammonical oxalate. There only remains to be found the means of preventing the destructive action which inevitably arises wherever copper and iron are in juxtaposition.

For the last ten years, the Société du Val d'Osne has supplied to the trade the most gigantic pieces which have ever been electro-coppered. We saw at the 1878 Exhibition two bulls ten feet long, weighing each 1600 kilogrammes (over $1\frac{1}{2}$ ton), which had been so coppered (in one piece), and were real masterpieces of manufacture.

WALENN'S PROCESS.—M. Walenn, a short time ago, made known in London the following process for coppering iron and cast iron:—

The preparation of pieces of metal, iron or cast iron, before being put in the bath, does not differ from what is done in other factories. An iron casting, for example, is at first scoured in dilute sulphuric acid, then rinsed in running water and dipped into a boiling solution of caustic potash. The piece so heated is put into the bath, which by its composition finishes the scouring should there be occasion for it.

As soon as the equilibrium of temperature is established the electric current is set on. The liquid used for the copper deposition is a mixture of potassic cyanide and neutral ammonic tartrate with a determined quantity of the metal in dissolution. The potassic cyanide maintains the metal in a state of dissolution and freely abandons it under the influence of the current. The neutral ammonic tartrate prevents the formation of precipitates. The bath, if suitably prepared, retains a normal composition without giving off any vapours or disagreable smells.

Walenn's process is characterised by the following advantages: No gas is liberated upon the piece when under treatment; the deposition is obtained with a comparatively small quantity of electricity at low tension; the same liquid is used whatever may be the thickness of the intended deposition; the duration of the operation is comparatively short; and lastly, the deposition is of a perfect homogeneity in all its parts.

This process being new cannot be fully appreciated at present, as the principal merit of a good deposition is to resist the action of time. All we now know is that M. Walenn has been coppering the plungers of the hydraulic presses used by the Thames Dock Company, and that this Company has declared itself perfectly satisfied.

COPPERING OF CAST-IRON ROLLERS FOR CALICO PRINTING. —Numerous trials have been made with a view of substituting cast-iron or iron rollers coated with a deposit of galvanic copper for the massive copper rollers previously used. Some had a comparative success in spite of the length of time required for the deposition; some others have been immediately abandoned owing to the bad quality of the deposited metal; actually, with the exception of the Wilde and Schlumberger's systems which are in operation in a few factories, there is no other process for coppering rollers which can be recommended.

WILDE'S PROCESS.—Mr. Wilde has, for a long time, made electrolytic deposition his speciality; his machines are in great favour in the North of England, and his competency in matters of electro depositions is universally acknowledged; his process for coppering has been in regular operation for some years at Manchester; it therefore deserves to receive the best attention at the hands of the practiser.

The basis of his system consists in giving to the cylinder which is being coated a rapid rotary motion in the copper bath, so as to constantly bring the metallic surface in contact with new layers of the electrolyte; this allows of the use of powerful electric currents for small surfaces of cathodes without interfering with the quality of the deposited copper.

SCHLUMBERGER'S PROCESS.*—The cylinder (having been perfectly cleaned by the usual methods) is made the cathode for the space of 24 hours, in a mixture of two liquids, composed of

	Parts.		Parts.
Water	12	Water	16
Cyanide of potassium ..	3	Sodic carbonate	4
		„ sulphate	2
		Cupric sulphate	1

It is then washed, rubbed with pumice powder, then washed again with an aqueous solution of cupric sulphate containing $\frac{1}{300}$ part of its volume of sulphuric acid, scraps of copper being kept in the bath to prevent the liquid becoming too acid. It is immersed again in the above alkaline solution, or else in a mixture composed of two liquids, viz :—

	Parts.		Parts.
Water	10	Water	10
Cyanide of potassium ..	3	Sodic carbonate	4
Aqueous ammonia	3	„ sulphate	2
		Cupric acetate	2

In these mixtures, at a temperature of 15° to 18° C., it is surrounded by porous cells containing zinc rods and dilute sulphuric acid, and connected with the zinc by copper wires. The cylinders are turned partly round once a day, in order to render the deposit uniform, and the action is continued during three or four weeks, until the deposit is $\frac{1}{25}$th of an inch.

COPPERING OF TELEGRAPHIC WIRES. — The industry of coating with copper steel wire used for telegraph lines is carried out on a large scale in the works of the Postal Telegraph Company at New York.

These works have 200 baths of sulphate of copper and 25 large dynamo-electric machines.

* 'Electro-metallurgy,' Gore.

The wire, which is wound on large drums, slowly passes through a series of baths, until it is coated with a sufficient thickness of copper. The duration of each operation is of about sixty hours. Sixteen kilometres of steel wire are covered in a day, the weight of the copper required being 250 kilogrammes. The total daily production of this immense factory may reach 50 kilometres.

COPPERING OF NON-METALLIC BODIES.—Hockin recommends, for coppering non-metallic bodies, dipping them into iodined collodion, afterwards immersing them in a solution of argentic nitrate and exposing them for a few seconds to the light; then into a bath of ferric protosulphate acidulated with nitric acid, and finally to deposit the copper on these bodies by means of an almost neutral solution of cupric sulphate.

CHAPTER XI.

ELECTROPLATING AND ELECTROTYPING.

Object of Electroplating—Simple Bath—Moulds in Plastic Materia's—Massive
Electroplating—Statuary or Relievo Electroplating—Qualities of Electroplating Copper—Electrotyping—Moulds—Baths—Modus Operandi—Duration of the Operation—Finishing of the Electrotypes—Mounting—Steel-
coated Electrotypes.

§ 1. ELECTROPLATING.

OBJECT OF ELECTROPLATING.—The invention of electroplating
dates from 1839, and has been claimed by Jacobi, Jourdan,
and Spencer. Jacobi, who also conceived the soluble anode,
having established his priority by the publication of his processes, is generally considered as the veritable inventor.

The sole object of Jacobi and his colleagues was the reproduction of medals and works of art by depositing on their
surface a non-adhering metal which could be used as a negative
mould. The operation was effected in a simple bath, and many
manufacturers are actually working by the simple bath process.

SIMPLE BATH.—This apparatus is composed, in principle,
of a wooden vat lined internally with lead or gutta-percha, and
containing a solution of cupric sulphate.

This vat (Fig. 27) contains some linen bags filled with
crystals of sulphate, for the purpose of maintaining the liquor
in a state of uniform saturation during the whole time of the
operation. In the middle are placed a certain number of
porous cells, generally made of unchilled porcelain, and containing dilute sulphuric acid and a zinc plate. All the zinc plates
Z Z Z are connected together by means of a copper rod T T',
the latter being itself connected with one or more smaller copper
rods, from which are suspended the pieces to be reproduced.

These pieces must be conductors; they act in the same manner as the copper electrode in the ordinary Daniell cell. The simple bath is nothing but a real Daniell cell closed on itself on a very short circuit, and in which the internal instead of the external current is used.

Fig. 27.

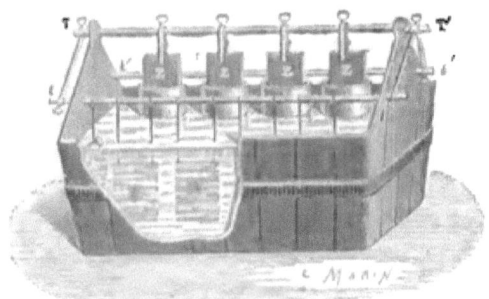

Simple Bath.

The bath is electrolysed as soon as the circuit is closed by the connection of the objects to the zincs; the zinc becomes oxidised, and the copper is slowly and in a very regular and uniform manner deposited upon the objects to be reproduced. When the coating of precipitated metal is sufficiently thick, the pieces are taken out of the bath and detached from the copper mould thus formed. These moulds are then placed in the bath and connected with the zinc. After a longer or shorter immersion in the cupric sulphate solution the moulds become coated with a deposit of copper, and perfectly true reproductions of the original objects are obtained.

It is easy to understand that the moulds of galvanic copper can only be used for taking impresses on objects having a sufficient taper, and that only objects which cannot be attacked by acid can be placed in the bath. If, for example, it is proposed to reproduce iron castings, it will be necessary to previously copper them, so as to prevent their being attacked by the acid solution. It is also necessary to rub the metallic surfaces with a pad soaked with turpentine, so as to prevent the adherence of the deposit against the pieces themselves.

MOULDS IN PLASTIC MATERIALS.—In practice gutta-percha is used for taking the impresses of metallic pieces, and gelatine for taking those of plaster or wax pieces.

The gutta-percha or gelatine moulds are rendered conductors either by the dry or by the wet method. The dry method consists in brushing the surfaces with plumbago until they become bright black. The wet method consists in coating the surfaces with a solution of argentic nitrate and alcohol, letting them dry, and blowing on them some sulphuretted hydrogen in a nascent state.

MASSIVE ELECTROPLATING.—In order to give the copper reproductions great strength without immoderately increasing the thickness of the deposition, Messrs. Christofle and Co. melt in the interior of the moulds some pieces of brass, which solder easily to each other, and give the electroplated depositions an appearance similar to that of castings. The pieces thus obtained can be worked exactly like cast copper; they possess the artistic appearance of the model, which has been faithfully preserved by electroplating, and, in the interior, a malleable resisting body.

RELIEVO ELECTROPLATING.—In the reproduction of statues or of large pieces in relievo the process consisted at first in executing the work in parts and joining these parts together by means of very careful solderings.

Mr. Lenoir has discovered the means of distributing and spreading the currents in the interior of a mould, whatever may be its dimensions, and therefore of obtaining very large pieces without soldering.

His process consists in replacing the single metallic wire of the negative pole of the battery by a platinum conductor made of numerous platinum wires interwoven with each other, so as to constitute a veritable carcase having *grosso modo* the shape of the interior of the mould with a little play, so as to leave everywhere a thickness of liquid between the walls and the platinum. The extreme wires are attached together, passing through a glass tube which isolates them from the gutta-percha mould, and are connected to the positive pole of an external battery.

M. Planté has substituted lead for platinum in the construc-

tion of the internal carcase, and this improvement has rendered
the Lenoir **process** perfectly practical.

In both cases, **with** platinum as **well as** with lead, care must
be taken **to leave an** opening **at the top, so as to** allow of the
escape of oxygen which **is constantly liberated round** the wires,
and a second opening **at the bottom to allow of** the liquid being
constantly renewed.

QUALITIES OF GALVANOPLASTIC COPPER.—The deposited
copper being chemically pure is not so alterable as the **merchantable** copper and alloys generally used in industry.

Its resisting **power** to tensile strain is greater by 20 **per 100**
than that **of cast copper.**

Its density is less than that of laminated copper and greater
than **that of cast copper,** the densities of these various materials
being 8·95, 8·75, and 8·85 respectively **for the** laminated, the
cast, and the galvanic copper.

§ 2. ELECTROTYPING.

ELECTROTYPES.—Electrotyping **comprises a series of** mechanical or electrical **means** for **reproducing** engravings **or**
typographical compositions.

The reproductions are called electros and electrotypes. **In**
order to produce an electro **the** original must **be** moulded **and**
the mould coated **with a** galvanic deposition; **a fusible metal
is** then run **at the back** of this deposition so as to strengthen it,
and the plate thus obtained is mounted on a piece of wood of a
determined thickness.

MOULDS.—The first operation **naturally** consists in taking
the impress of the engraving to be duplicated. Gutta-percha,
or impermeable plaster, or one of the following mixtures **may**
be used **for the purpose** :—

I.	White wax	..	..	..	..	200	grammes.
	Spermaceti	..	..	..	..	30	,,
	Stearine	..	..	..	..	250	,,
	Plumbic carbonate	..	..	..	30	,,
II.	Glue	..	..	..	..	400	,,
	Treacle	..	..	..	..	100	,,

This mixture gives some elasticity to the mould.

III. Bismuth ... . 250 grammes.
   Lead 160   ,,
   Tin 125   ,,
   Antimony 30   ,,

Mix, and melt in a clean crucible.

IV. Bismuth .. .. 280 grammes.
   Lead 190   ,,
   Tin 100   ,,

In order to obtain a good result from the two last formulæ the metals are first melted and poured in a vessel containing cold water and a small quantity of straw and hay cut in lengths of about 3 inches. The whole is thoroughly stirred whilst the molten metal is being poured. This divides the metal into shots, which are dried and melted again.

V. Gelatine .. .. 500 grammes.
   Water 700   ,,
   Wax 15   ,,

The gelatine is dissolved in water on a gentle fire and some beeswax in small pieces is added. This mixture must be used tepid and not hot.

VI. Beeswax .. .. 9 kilogrammes.
   Venetian turpentine .. .. .. 1·35   ,,
   Plumbago in an impalpable state 0·225   ,,

Care must be taken to avoid any dust. If during the cold weather any crack occurs a little Venetian turpentine is added. When the temperature is sufficiently high turpentine can be dispensed with altogether.

According to Mr. Urquhart, this last mixture is extensively used in England.

The wood engraving is first cleansed with a hard brush and essence, then dried in sawdust, and sprinkled with extremely fine plumbago. Some wax of good quality is then melted into a copper pan and slowly stirred at the same time that the plumbago is being poured in. The mixture must be thorough and pasty, almost liquid. It is necessary to maintain it at a certain temperature without boiling it, and to stir it well to exclude all moisture and air-bubbles. The wax is then poured into shallow metal vessels about one quarter inch deep. (These vessels are provided with rings at the angles to facilitate their being suspended in the bath.)

The mixture of moulding wax, when very hot, is poured in the shallow metal vessel so as to spread uniformly, and the said vessel is placed on a horizontal surface to let the wax partially cool. The surface must be skimmed as soon as cracks are observed and a skin begins to form. When the wax is in a kneading state it is placed on the engraving and slightly pressed. The mould is then raised a little in order to ascertain if any particle of wax has remained attached to the wood, in which case the latter is again sprinkled with plumbago and placed back in its former position. Precaution must be taken by means of guiding marks or otherwise to secure the replacing of the mould in exactly the same position which it previously occupied.

A strong pressure is then applied, either by means of a hydraulic press (Fig. 28), or of a hand-screw press. The

FIG. 28.

Hydraulic Press for Moulding the Electros.

hydraulic press devised for electrotyping is small and compact. The table can be withdrawn so as to enable the workman to well secure his mould in position. The pump is placed at one of the sides. A graduated scale constantly indicates the energy of the pressure.

The screw press is perhaps less liable to accidentally get out of order than the hydraulic, but its action is less intense and less uniform.

When this operation is done, the mould is lifted and the wax is examined; it must represent a faithful sunk copy of the engraving. The impress is sprinkled with fine plumbago and polished with a very smooth brush.

In America they employ for plumbago coating purposes a machine (Fig. 29), the use of which is more economical and healthy. When working by hand the dust flies everywhere,

Fig. 29.

Plumbago Coating Machine.

whereas the machine is so arranged that the brushes almost cover the mould, and the excess of plumbago drops into a lower vessel. The construction and working of this machine are most simple; it is a horizontal frame mounted on four legs, with a sliding table composed of a series of bars. A long brush of the same width as the table receives, from a spindle rotating at 400 revolutions per minute, a vibratory motion. The machine may be covered during the operation, so as to avoid breathing the plumbago dust.

The brushes are made of goat's hair of the finest quality.

The plumbago must be very pure and well crushed, and, before being used, passed through a very fine sieve. It is essential to take great precautions in the preparation and use of the plumbago, for almost all the imperfections of the electros arise from a neglect in the metalisation of the surfaces which are to receive the galvanic copper deposition.

The machine gives a regular work and avoids the numerous spots which are often to be seen on the electros the moulds of which have been plumbago coated by hand.

BATHS.—Simple baths, that is to say, electric cells in which the moulds are immersed, are still in extensive use, but the really industrial process consists in using depositing baths and sending through them the current produced by a dynamo-electric machine. In the first instance the liquid used in the cell is a nearly saturated solution of cupric sulphate, with 120 grammes of sulphuric acid for each 10 litres of liquid. To obtain a better deposition, 4 grammes of arsenious acid are added, or, instead, a small quantity of stannic chloride.

The cupric sulphate must be dissolved in distilled boiling water; it is left to cool, and the sulphuric acid is then added. The cupric sulphate must be very pure.

In the second case, when the electric current is produced outside the bath, the liquid used must be composed as follows:—

Sulphate of copper	825 grammes.
Sulphuric acid	825 ,,
Distilled water	10 litres.

The water is first put in the vat and the acid is added by small quantities at a time, stirring constantly; then a quantity of sulphate of copper as great as the acidulated water will dissolve cold is put into the solution, which is maintained in a state of agitation.

The saturated bath must be at 25° Baumé; it is used cold and must be maintained saturated by the addition of crystals.

The sizes of the vats are naturally in proportion to the

importance of the work in hand. We will only make upon this
subject a single remark, that is, that their depth should be
sufficient for lodging not only the electrodes but also the
sediments resulting from a long period of working. The copper
solution being denser at the bottom than at the top, it is
necessary, in order to obtain a regular thickness of deposit on
the mould, to continuously stir the liquid.

If the wooden vats can be made water-tight without any
internal lining, this would be better than any other system.
Otherwise we would recommend a lining made of sheet lead
soldered together with a red soldering iron in preference to a
lining of gutta-percha, marine glue, or any other substance.

MODUS OPERANDI.—When the mould has been coated with
plumbago, all the parts which are to remain intact are touched
with a brush dipped in hot wax, and it is then immersed in
slightly alcoholised water, a jet of water being afterwards
directed upon it so as to remove the excess of plumbago and
the air bubbles which adhere to the surface.

The mould is afterwards placed in the bath; a very small
surface of anode must at first be immersed to prevent too rapid
a deposition; this is important, as otherwise the copper might
be deposited in a granulated state. The current must be
intelligently regulated; it should be moderated at the beginning
or the deposit would have a granulated appearance and a black
colour. When the deposit has commenced in good conditions
a bright layer starts from the suspending wire and extends
itself gradually radiating upon the surface coated with plumbago. After a short time the anode is gradually lowered down,
thereby increasing the surface of action according to the progress of the deposition. When the mould is completely covered
the anode can be completely immersed and the strength of the
current increased without inconvenience.

The principal consideration for a manufacturer is to produce
well and rapidly. Certain experimenters, guided by a long
experience and the exact knowledge of the principles of their
art, work twice as quickly as some others, obtaining at the same
time copper electros with a very fine and very close grain.

Many causes may interfere with the rapidity of the deposition, even when the intensity of the current allows of operating quickly. If the bath is too dense owing to an excess of copper, the deposit will form slowly and have a crystalline appearance. If the bath is too poor, the deposit may be quickly effected but the grain will be porous, which will be still worse. Variations of temperature also exert a considerable influence upon the speed of the deposition and the quality of the metal. It is necessary, as far as possible, to maintain in every season the temperature of the bath at 16° C. The purity of the anodes, their surface, and their distance from the moulds, equally constitute the essential elements of a good result.

The anodes must be of chemically pure copper, which is now very easy to obtain owing to the progress accomplished by electrolysis in the refining of copper. Impure anodes are extremely annoying as they get covered with certain impurities which are insoluble in the bath. Their surface must be at least equal to those of the moulds to be coppered; too small an anode weakens the solution, and too large a one enriches it. Their distance from the moulds varies with the dimensions of these moulds and the intensity of the current; it generally varies from 2 to 5 centimetres.

DURATION OF THE OPERATION.—According to Mr. Urquhart a deposition of copper of 3 millimetres' thickness can easily be deposited in two days, and if the work is pushed with all the available power of current a good electro with fine close grain can be obtained in ten or twelve hours. (A single coppering can be effected in one hour at a maximum.)

According to M. Stoesser the average duration of an electrotyping operation is twenty-four hours. The average thickness of the deposition is ·3 millimetre; it corresponds to a layer of 25 grammes per square decimetre or a deposition of about 1 gramme per hour per square decimetre. The author states that the intensity of the current may be doubled, and a deposition of the same quality be obtained in twelve hours, and in this he is in accordance with Mr. Urquhart, but the duration of the twenty-four hours' operation is convenient for preparing the moulds during daytime, and putting them in the bath in the evening.

It is, however, useful, and particularly to the large illustrated papers which require prompt illustrations of the most recent events, to obtain electrotypes in less than twelve hours, and good proofs have been obtained in eight, six, and even four hours ; but this should not serve as a basis for a normal production in an industrial workshop, as the electros become more expensive and are not so regular.

The best means for operating with the greatest possible speed is to coat the mould, before its immersion in the galvanic bath, with a concentrated solution of sulphate of copper, and to spread over it some iron filings, perfectly pure and without grease. The filings and the sulphate are mixed with a camel's hair brush. The iron absorbs the acid, and the copper is uniformly deposited on the mould. The mould is then washed with a jet of water and immersed in the bath.

Adams recommended spreading fine tin powder on the wax when still hot ; but notwithstanding its increased conductivity, the mould could not receive too great a number of amperes per square metre, and the operation was not much quickened.

FINISHING OF THE ELECTROTYPES.—When the operation is terminated, the frame is removed, the electrotype washed in running water, and the edges which could prevent the deposit being removed are chipped. The frame is heated over a gas jet so as to detach the wax and liberate the electrotype. The latter is then placed on an iron slab, brushed externally with hydrochloric acid saturated with zinc (spirits of salts for soldering), and sprinkled with small pieces of solder.

The slab is placed in a cell filled with molten stereotyping metal, until the solder melts and it can be spread with a rag or some oakum. This operation is called the tinning of the electro. Finally the slab is taken away and a frame is formed round it with some iron bars, and a sufficient quantity of metal is run into the frame. This metal is generally composed of 91 parts of lead, 5 of antimony, and 4 of tin.

The thickness of the metal backing the electro varies between 4 and 10 millimetres according to the size of the engravings.

MOUNTING.—There only remains to mount the plates on

wooden blocks. The edges are cut with a circular saw, and the profiles of the metal are planed according to a determined caliper. The borders are obliquely filed so as to disengage the top part, and the plate is mounted on a piece of oak or mahogany.

Before effecting the mounting the electro must be carefully examined, and should there be any depression found it should be straightened by slightly tapping it at the back.

The machine (Fig. 30), the working of which can be understood by a mere reference to the illustration, is intended for

Fig. 30.

Machine for regulating the Thickness of Plates.

reducing the plates to very regular thicknesses, and the machine (Fig. 31) for shaping the wood backings on all their faces. These two machines are indispensable for obtaining good electros, which are easily brought up in the printing machine or press.

STEEL-COATED ELECTROTYPES.—When an electro has to be used a great number of times it is useful to cover it with an iron pellicle. By this means the duration of the electro is prolonged, and a more artistic impression obtained than with the copper only. The deposit must be exceedingly thin, otherwise

N 2

the lines might come thick, which should be particularly
avoided.

Before beginning the deposition of iron, it is necessary to
perfectly clean the electro and rub it after with caustic potash.

Finishing Machine for Electros.

The bath is principally composed of ammonic carbonate in
the proportion of 12 kilogrammes of carbonate to 75 litres of
water. The iron must be dissolved in the solution by immersing
an anode of charcoal iron connected with the positive pole of a
3 or 4 Bunsen cell battery. A second iron plate is connected
to the negative pole, and the deposition is from time to time
tested by replacing the cathode by a copper plate.

With one or the other of these baths, the operation is conducted as follows:—

An anode of the best quality of charcoal iron is used, and a
current of about 4 volts is sent through the bath. The deposition is not immediate as is the case with copper: the electro
to be steel coated must remain immersed during a few minutes;
after four or five immersions the deposit will be of a sufficient
thickness.

The plates thus coated must, as soon as taken out of the

bath, be carefully washed in boiling water.   They are afterwards washed and brushed in cold water, and then dried and **rubbed** with turpentine.   If they **are** not immediately used **a** layer **of** melted **wax** should be **spread on** their surface.

Messrs. Christofle **& Co.** exhibited **in** 1881, in Paris, some **very curious electros on which** the **steeling had** been replaced **by** nickeling, **not** by the surface nickeling, **which** would have **had the** inconvenience of obliterating the lines of the engraving, but by **direct** galvanic reproduction of nickel rendered thicker by the galvanic backing of the electro.   These electros were obtained from a gutta-percha mould, **on** which **a** slight coating of **nickel,** strengthened by **a** copper backing, had **been** deposited; **they were mounted** like ordinary **copper** galvanic electrotypes.

CHAPTER XII.

DEPOSITION OF VARIOUS METALS.

Deposition of Copper, Zinc, and Tin Alloys—Deposition of Platinum—Deposition of Lead—Deposition of Zinc—Deposition of Iron—Deposition of Tin—Tin Crystals—Deposition of Antimony—Deposition of Aluminium—Deposition of Cadmium—Deposition of Cobalt.

DEPOSITION OF COPPER, ZINC, AND TIN ALLOYS.

SALZÈDE'S PROCESS.—The baths used by M. Salzède are composed as follows:—

1. *Brass Bath.*

Potassic cyanide	12 parts.
Potassic carbonate	600 „
Zinc sulphate	48 „
Cupric chloride	25 „
Ammonic nitrate	305 „
Water	5000 „

Dissolve the cyanide in 120 parts of water; dissolve the potassic carbonate, the zinc sulphate and the cupric chloride in the remainder of the water, raising the temperature to about 60° C. When the solution is made, add the ammonic nitrate, stirring the mixture. Let the liquor set for a few days, and use it after decanting, when no more sediments are formed.

2. *Bronze Bath.*

Potassic cyanide	50 parts.
Potassic carbonate	500 „
Tin chloride	12 „
Cupric chloride	15 „
Water	5000 „

This bath is used at a temperature not exceeding 36° C.

RUSSELL AND WOOLRICH PROCESSES.—Messrs. Russell and Woolrich recommend the following brass bath :—

Brass Bath.

Cupric acetate	3730 grammes.
Zinc acetate	370 ,,
Potassic acetate	3730 ,,

Dissolve the above substances in 25 litres of hot water, and add some cyanide until the formation of a precipitate, which is dissolved in an excess of reagent. Maintain an excess of cyanide in the bath. Use either one brass anode, or at the same time, two anodes, one of brass, the other of copper.

BRUNEL'S PROCESS.—M. Brunel prepares his solution in the following manner :—

Brass Bath.

Cupric chloride	373 grammes.
Potassic carbonate	9325 ,,
Zinc sulphate	746 ,,
Ammonic nitrate	4660 ,,

Dissolve the chloride in 2·25 litres of water, the carbonate in 27 litres of water, the zinc sulphate in 2·25 litres of hot water. Mix these solutions. Add the ammonic nitrate, and mix again very intimately. Add some cold water to make 50 litres.

Another formula :—

Potassic carbonate	3730 grammes.
Potassic cyanide	570 ,,
Zinc sulphate	466 ,,
Cupric chloride	310 litres.
Water	56 ,,

Prepare separately the various solutions. Add to the zinc sulphate and to the cupric chloride a part of the dissolved carbonate. Dissolve the precipitates which are formed by adding ammonia; introduce the remainder of the potassic carbonate and the cyanide, and complete with water.

NEWTON'S PROCESS.—Mr. Newton has prepared various solutions for obtaining the deposition of copper, tin, and zinc alloys.

His method is based on the use of mixtures of zinc chloride with alkaline chlorides, potassic, sodic, ammonic; of zinc acetates, with the corresponding alkaline acetates.

For brass baths, use copper with the same acids as the salts of zinc, acetates for example.

For bronze baths, dissolve the double tartrate of protoxide of tin and potassium, with or without addition of caustic potash.

Mr. Newton deposits an alloy of zinc, tin, and copper by means of a bath composed of double cyanide of copper and potassium, potassic zincate, and stannate; he obtains these latter by melting tin oxide with caustic potash, or by dissolving in potash.

For brass baths he uses a solution composed of cupric oxide dissolved in an excess of potassic cyanide, zinc oxide, a small addition of liquid ammonia, and heats to about 50° C. The solution is completed with water, so that the liquid contains 2 of zinc for 1 of copper.

Other Baths.

Potassic cyanide	373 grammes.
Ammonic carbonate	373 „
Cupric cyanide	62 „
Zinc cyanide	31 „

Dissolve in 4·50 litres of water. Raise the temperature to 65° C.

Potassic cyanide	Equal weights.
Ammonic carbonate	

Dissolve in 4·50 litres of water. Attach a large brass anode to the positive conductor of the battery, and use a small surface of cathode, a thin sheet of brass for example. The temperature must be about 65° C. The anode is dissolved, and furnishes the metal required for the bath.

DEPOSITION OF PLATINUM.

ROSELEUR'S PROCESS.—M. Roseleur was the first to succeed in obtaining good platinum deposits. His patent is dated 1847.

The following is the process which he indicates as suitable for depositing on copper or its alloys any quantity of platinum.

Introduce in a long-neck glass matrass 10 grammes of finely laminated, or even better, frothy or spongy platinum, with 150 grammes of hydrochloric acid, and 100 grammes of nitric acid at 40 degrees. Heat on a piece of sheet iron pierced with a central hole so that only the bottom of the matrass receives the impression of the heat. Large volumes of rutilant vapours are liberated, and the platinum entirely disappears, leaving only a red liquid, which must be continued to be heated until it becomes so viscous that it will stick to the walls of the matrass. The last part of this operation can be effected in a porcelain cup, the truncated form of which is more favourable to the evaporation of liquids in excess. It is then taken out of the fire and allowed to completely cool, when it is afterwards dissolved in 500 grammes of distilled water, and filtered if required.

At the same time 100 grammes of ammonic phosphate is dissolved in a similar quantity of distilled water, and the two solutions are mixed. An abundant precipitate of ammoniaco-platinic phosphate is formed, upon which floats an orange-coloured liquid which must not be separated.

When in that state, a solution, prepared in advance, of 500 grammes of sodic sulphate in one litre of pure water is slowly poured, the solution being continually stirred.

The mixture is then brought and maintained at the boiling point, the evaporated water being replaced, until, owing to the escape of ammonia, which can be detected by the smell, the liquor from alkaline turns sensibly acid to the litmus paper. In this reaction the liquor, from yellow, turns perfectly colourless, which is indicative of the formation of the double salt of platinum.

The solution is then ready to operate on copper and its alloys by means of heat and a moderately powerful electric current.

AMERICAN PROCESS.—Platinic chloride is prepared as above described, and dissolved in distilled water; potassic cyanide is added, an excess of which will redissolve the precipitate formed.

The solution must contain about 65 grammes of metal for 10 litres of water.

The current must not be too strong, otherwise the deposition of the platinum would be effected in a state of black powder.

The platinum anode not being attacked by the cyanide, the bath grows weaker, and its strength must be maintained by adding some platinic chloride from time to time.

DEPOSITION OF LEAD.

FRENCH SOLUTION.—Dissolve 10 grammes of litharge (protoxide of lead) in 100 grammes of caustic potash for 2 litres of distilled water. Use a lead anode, and add some litharge from time to time.

AMERICAN SOLUTION.—Simply dissolve plumbic acetate or nitrate in water. The lead is easily deposited when the solution is only slightly concentrated and the current weak. For laboratory experiments an *alkaline* bath is used; it is prepared by precipitating the lead of plumbic acetate or nitrate by potash, soda, or ammonia, and dissolving the precipitate in potassic cyanide.

CHROMOPLASTICS, OR COLOURED RINGS.—Beautiful colorations may be obtained with a battery or a Gramme machine, on polished steel plates by means of preparations of lead salts. M. Roseleur gives the following process:—"The object to be coloured, attached to the negative pole, is immersed in a cooled bath of plumbite of soda; a platinum anode is then gradually dipped into the liquor without touching the object with it. The latter is immediately observed to become coloured with a variety of hues : yellow at first, and each of these colours grow darker and completely change as the platinum anode is more or less deeply immersed. These lighting effects, which are caused by the greater or lesser thicknesses of the deposited plumbic acid, can be indefinitely varied.

Mr. Watt proceeds in the same manner, but uses plumbic acetate instead of protoxide.

If a piece of sheet copper cut in a star shape is attached

to the anode, the colorations on this plate are effected in an analogous manner. If a piece of cardboard is placed between the conducting wire and the plate, the colorations take place outside this improvised screen.

The colours so produced resemble those of the prism, but as they do not firmly adhere to the metal, the latter must be carefully washed with distilled boiling water, and, when dry, coated with a layer of good white varnish with alcohol.

DEPOSITION OF ZINC.

The process of depositing zinc by simple immersion on well scoured iron is generally designated by the name of galvanisation. This deposit, which has nothing galvanic besides its name, protects iron from oxidation for a comparatively long time.

Many attempts have been made at obtaining good deposits with batteries, but we do not believe that any industrial results have ever been obtained.

The following is a process, somewhat old, described by Mr. Watt :—

Dissolve 7 kilogrammes of commercial potassic cyanide in 100 litres of distilled water. Pour into the solution 2·50 kilogrammes of concentrated ammonia (density ·880). After stirring and mixing well, place a few porous cells, such as Daniell's, in the solution, and fill them with a concentrated solution of potassic cyanide (about 550 grammes for 5 litres of water) until the levels inside and outside are even. Place in the porous cells some pieces of iron or of copper connected with the negative pole of a battery. Place in the solution of potassic cyanide and ammonia some pieces of laminated zinc of good quality, and scoured with hydrochloric acid, and connect them with the positive pole of the battery. Send the current through until the solution of cyanide and ammonia has absorbed about 2 kilogrammes of zinc, that is to say, 10 grammes per litre. Then remove the zines and the porous cells, dissolve 2·5 kilogrammes of alkaline carbonate (potash preferably) in a portion of the preceding solution, and mix the whole, stirring and mixing with care. Let the bath settle, and decant.

Use the bath with laminated zinc electrodes. For plane-surface pieces place a zinc sheet on each side: thus, for sheet iron, place alternately one sheet of zinc and one of iron, and so on.

DEPOSITION OF IRON.[*]

Iron has a very great chemical analogy with nickel, and the majority of remarks made on nickeling apply to iron deposition. The solutions are corresponding; iron taking the place of nickel; but, owing to the tendency of iron salts to acquire a greater degree of oxidisation, the iron solutions become rapidly altered. The hydrogen liberated by electrolysis goes to the cathode, and the corresponding oxygen combines itself with the iron anode, and has a tendency to produce basic salts. For these reasons the double chloride of iron and ammonium is much more advantageous than the sulphate.

The double chloride of iron and ammonium is prepared by dissolving unoxidised iron wire in hydrochloric acid, heating at the end of the operation, and having an excess of iron to prevent the formation of perchloride. For 60 grammes of dissolved iron 55 grammes of ammonic chloride are added to the solution. Mr. Sprague has found that the alteration of the solution was retarded by the addition of a proportion of glycerine.

A solution of iron altered either by the action of the air or by the generation of basic salts, can be renovated by adding to it the suitable acid, heating the liquor until it is clear, and sending through it an electric current for dissolving an iron anode.

The iron deposition may be used for important scientific experiments as regards the laws on magnetism, but it does not seem to us to admit of lucrative industrial applications. Its principal usefulness appears to consist in the steeling of electro-types, which we have previously described.

DEPOSITION OF TIN.

Galvanic deposition of tin is not much in use, as tinning by simple immersion or by contact with a second metal generally

---

[*] Sprague, 1884 edition.

affords more economical results; in some cases, however, electricity must be preferred as giving more adhering and durable deposits.

Amongst the numerous electro-tinning **processes** tried **with** more or less success, we will mention those **of Messrs.** Roseleur, Fearn, Lobstein, Maistrasse, and Birgham, and we will very briefly analyse them.

Roseleur's Process. — The **bath** recommended **by M.** Roseleur is composed as follows :—

Distilled water	50 litres.
Pyrophosphate of sodium	500 grammes.
Melted stannous protochloride	50 ,,

The **water is** contained in a **vat entirely lined with sheet tin** or tin **anodes; the** pyrophosphate is **poured in and** the mixture **stirred until it is dissolved. The stannic** protochloride is **introduced, on a brass sieve, in the midst of the** bath, and the solution **is again** stirred **until a** complete dissolution **is** arrived **at. The** liquid **then** becomes clear **and almost** colourless.

The anodes are not sufficient **for keeping the solution in a** state of saturation; **it is necessary, when the deposition gets** slower, to add, by **small quantities at a time, some tin salts and** some sodic pyrophosphate.

Fearn's Process.—This process, which **is used by the** Electro-Stannous Company of Birmingham, comprises four different solutions, two for thick **and** two for thin deposits.

The following are the two in use for obtaining thick deposits :

1st. A solution of stannous chloride containing 100 grammes of tin per 5 litres is first prepared; then a solution of 15 kilogrammes of caustic potash in 100 litres of **water;** and lastly a third solution of 15 kilogrammes **of** sodic pyrophosphate in 30 litres **of water.**

The liquid potash is poured into the tin solution, continually stirring with a glass rod; 15 kilogrammes of potassic cyanide are then added, **and** lastly the 45 kilogrammes composing the solution of sodic pyrophosphate, keeping the mixture in a **constant** state of agitation.

2nd. Prepare separately :

A solution of 12 kilogrammes of potassic tartrate in 320 litres of water.

A solution of 34 kilogrammes of caustic potash in 230 litres of water.

A solution of 20 kilogrammes of stannic chloride (100 grammes of tin per 5 litres).

These three solutions are mixed as previously indicated, stirring continuously.

The electric current must have a tension of 3 to 4 volts.

The iron and cast iron articles are covered with a copper coating before being tinned.

LOBSTEIN'S PROCESS.—We find in Mr. Napier's treatise the following information relating to M. Lobstein's process which has been more particularly used in England. The battery is formed with a leaden vessel $1 \cdot 20$ metre long by $\cdot 60$ metre wide and $2 \cdot 25$ metres deep. At the bottom are placed a copper sheet of $1 \cdot 35$ metre by $0 \cdot 70$ metre, on the top of which is a zinc sheet of $1 \cdot 05$ metre by $\cdot 67$ metre. The exciting solution contains 4 kilogrammes of plumbic acetate, 4 kilogrammes of common salt, and 1 kilogramme of sulphuric acid in as much water as may be required for filling the vessel until the copper and zinc are covered.

The tinning bath is $2 \cdot 40$ metres long by $1 \cdot 20$ metre wide, and $1 \cdot 20$ metre deep. It contains 2270 litres of water in which has been dissolved 40 kilogrammes of caustic soda, 1 kilogramme of potassic cyanide, and 700 grammes of tin salts. The tin electrodes are connected to the copper of the battery; the iron sheets upon which the deposit is effected measure 15 square metres.

The result of an operation lasting 96 hours was:

Zinc dissolved in the cell	748 grammes.
Tin dissolved in the electrodes	888 „
Tin deposited on the iron	361 „

As may be seen, this method is neither rapid nor economical; we describe it as it received a few years ago a great publicity.

MAISTRASSE'S PROCESS.*—M. Maistrasse uses a tinning bath

* Julius Weiss, 1883.

containing 1000 litres of a solution of caustic soda concentrated
at 3° Baumé; 100 grammes of stannic chloride and 300 grammes
of potassic cyanide. He uses anodes of pure tin which are
dissolved during the operation.

The cast-iron objects, well pickled, are immersed during
24 hours in this bath, and kept in communication with a large
surface Daniell cell.

This experimenter heats the tinned objects to the temperature of fusion of tin, which gives them an exceptional
appearance and solidity.

Tinned zinc heated under these conditions becomes more ex-
tensible and more easy to laminate and solder than ordinary zinc.

BIRGHAM'S PROCESS.—Mr. Birgham tins all the metals by
dissolving merchantable tin in hydrochloric acid, precipitating
the solution by means of a lixivium of potash and mixing it
with a second solution of caustic potash and potassic cyanide.

OBTAINING CRYSTALS OF TIN BY MEANS OF ELECTROLYSIS.

The crystallisation of tin offers a remarkable phenomenon
when a solution of stannous chloride is electrolysed. The
crystals of tin formed upon the cathode increase so rapidly in
length as to grow across the solution and touch the anode in a
few minutes. If the solution is concentrated, the current
strong and the cathode small, the mass of crystals rapidly fills
the whole vat, converging towards the anode. If the anode is
taken further apart the crystals are seen to follow it.

In order to obtain large size crystals a platinum cup is
covered externally with a layer of wax leaving the bottom
exposed; this cup is placed on a sheet of amalgamated zinc
contained in a porcelain vessel. The cup is afterwards entirely filled with a dilute solution, not too much acid, of stannic
chloride, and the external vessel is filled with hydrochloric
acid diluted with nineteen volumes of water up to the height
at which the two liquids come into contact. The energetic
current generated reduces the tin salt; in a few days' time the
crystals are much developed. They must be washed in large
volumes of water and quickly dried.

DEPOSITION OF ANTIMONY.

The galvanic deposition of antimony having been specially studied by Mr. Gore, we will borrow from him the description of the processes employed.

Antimony may be deposited by simple immersion, and by means of an electric current; in the latter case the metal may not only be obtained in a state of loose black powder, but also in two distinctly different coherent reguline conditions, viz. as a very brittle metal of a grey slate-colour and hard crystalline structure; and also in a highly lustrous steel-black deposit, of amorphous structure.

The solution used for obtaining the pure grey metal is composed of :—

Distilled water	350 grammes.
Tartar-emetic	30 ,,
Tartaric acid	30 ,,
Pure hydrochloric acid	45 ,,

It is not a good conductor, and should be used with a current of about 1 volt, so as to deposit about 1 millimetre per week.

Mr. Gore describes a curious phenomenon which came under his observation when depositing antimony. " If," he says, " at any moment during the formation of the deposit the piece is taken out of the bath and gently tapped or rubbed with a hard substance, an explosion is produced accompanied by a slight cloud of white vapours and sometimes with light, and almost always evolving an intense heat. If the deposited antimony is not homogeneous, the explosion attacks the subjacent metal to a depth of 3 millimetres. Similar phenomena sometimes occur in the course of the electrolytic operation, when the deposit rubs against the walls of the glass cell which contains the solution."

For obtaining a bright shining deposit, the following solution can be used :—

Sulphate of antimony	500 grammes.
Potassic carbonate	1 kilogramme.
Water	8 litres.

It is boiled, and when hot submitted to the action of an electric current with an antimony anode.

We do not insist on the deposition of this metal, for it is very little used in industry.

DEPOSITION OF ALUMINIUM.

M. A. Bertrand deposits aluminium by means of a solution of double chloride of aluminium and ammonium. He asserts that he obtains on a plate of copper a brilliant deposit capable of receiving a beautiful polish. But his process, when tried by a few chemists, has given nothing but results insignificant as to quantities and disastrous as to quality.

Mr. Urquhart obtains white deposits by using concentrated sulphate of aluminium acidulated with a little sulphuric acid. The temperature is about 65° C. The electromotive force varies from 3 to 8 volts. Mr. Sprague declares that it is impossible to deposit aluminium, and one would feel inclined to share his opinion after perusing the specification of the patents taken out upon the subject as well as the notes communicated from time to time to the scientific publications by clever experimenters. In any case, the deposition of aluminium is not industrially practised, and has therefore only a secondary interest for us, since we are considering the practical side of the question alone.

DEPOSITION OF CADMIUM.

According to Smee it is difficult to obtain firm and coherent deposits of cadmium from solutions of either its chloride or sulphate; but it can easily be deposited from a solution of the ammonio-sulphate, prepared by adding sufficient aqueous ammonia to a solution of sulphate of cadmium to redissolve the precipitate.

Napier recommends the following method:—

A solution of cadmium is easily prepared by dissolving the metal in weak nitric acid, and precipitating it with carbonate of soda; then washing the precipitate and dissolving it in potassic cyanide.

o

This solution must be heated to 40° C. and electrolysed by a current of 4 to 6 volts. The deposited metal is white, and resembles tin. It is very soft and does not offer any important advantage in industry.

DEPOSITION OF COBALT.

A concentrated solution of the chloride, with its excess of acid neutralised by the addition of caustic potash or ammonia, and electrolysed by a weak current, deposits its metal in uniform layers. The deposited metal is brilliantly white, hard and brittle, and can be obtained in cylinders, bars, and medals by using proper moulds to receive it. The deposited rods are magnetic.

With an anode of cobalt it is unnecessary to alter the solution after its preparation. Rods of cobalt electrolytically deposited contain as much as thirty-five times their volume of hydrogen.

DEPOSITION OF GERMAN SILVER.

German silver is an alloy of copper, zinc, and nickel.

A suitable bath is prepared by dissolving German silver of good quality in nitric acid, and adding to the solution a solution of potassic cyanide, stirring slowly until all the metal is precipitated. The exhausted liquid is then poured out, the precipitate is washed, redissolved in potassic cyanide, and diluted with double its volume of water.

Or 1 kilogramme of cyanide of potassium and an equal quantity of carbonate of ammonium may be dissolved in 10 litres of water, the solution being heated to 70° C., and the German silver dissolved in it by means of the electric current. If the deposit becomes too red, add carbonate of ammonium, if too white, add cyanide.

All the alloys of this description require very strong currents to be effectively decomposed.

SECTION IV.

CHAPTER XIII.

REFINING OF COPPER AND LEAD.

Amongst the various metals, the refining of which is obtained by means of electricity, we shall mention copper and lead: copper in various English and Continental factories, lead in the Electro-metal Refining Co.'s works, of New York.

COPPER REFINING.—To obtain chemically pure copper, and extract from raw coppers the often considerable quantities of gold and silver which they contain, a bath of sulphate of copper and thin copper cathodes, previously purified, is often used.

The electric current must here, as in all applications of electrolysis, supply the work required by the chemical decomposition, and overcome the various resistances which oppose its passage in the bath: metallic resistance, polarisation of electrodes, &c. The electric energy must also effect the deposition of the metallic atoms on the cathodes.

In a bath of sulphate of copper, with copper anodes, the sulphate is decomposed by the action of the current, and the copper is deposited on the cathode; the sulphuric acid attacks the anode and regenerates the sulphate of copper decomposed by the current, so that the work performed is equal to the work expended.

If the metals operated upon had no impurities, and if the operation were conducted with care, no gaseous escape would take place, and there would consequently be no polarisation.

In electro-plating, the work performed in the carriage of particles may be considered as *nil*, owing to the fact that the resistance of the baths for the weight of deposited metal is insignificant; but in refining work, when a single machine often precipitates more than 10 kilogrammes of copper per hour, the motive power necessitated by the said carriage is not to be neglected. This part of the expenditure is, however, always very small, and the electrical energy is almost entirely consumed in the overcoming of the resistance offered by the baths to the passage of the current.

It is, therefore, sufficient to know the intensity of the current in amperes and the resistance of the baths in ohms in order to calculate the power required for a given production of copper. A coefficient, varying with the nature of the coppers under treatment and the speed of the machine, is afterwards applied in each particular case.

Calling this coefficient k, the formula expressive of the motive power will be

$$W = k \frac{R\,C^2}{g} \text{ kilogrammetres.}$$

But it is always possible to reduce the resistance of a bath by increasing the surfaces of the anodes and cathodes, and to reduce the intensity of the current for a given production by joining the baths in series and giving them a total resistance equal to that of the primitive bath. It is, therefore, possible to at will reduce the expenditure of electricity, and consequently that of the motive power, and this in a very large proportion. For example, it is easy to conceive a factory with a plant laid down for the refining of 1 ton of copper per hour, with the expenditure of only 1 horse-power, if the economy of motive power is contemplated **exclusively of the** space required, **the** cost of installation, and **the real cost price** of the refining.

The expenditure **of power is, in fact,** only one of **the** economical **elements of the question,** and when waterfalls are available it is **often** the **least** important **one.** The sizes of the installations and the quantity of metal to be **treated** may be the determining **causes** of the failure of a manufactory; for **the** interest on the capital absorbed may reach, **and even exceed** the gross profits realised by the operation itself. **When a** quantity of copper is refined with a given power, **and it is pro-** posed to double the production without increasing the expenditure of power, **it** becomes necessary to quadruple the quantity of metal **to be** treated, which increases in a very great proportion the **cost** of first **establishment.** The absorbed capital then **becomes** considerable **in proportion to** the annual amount of **business.**

The limit beyond which **it becomes** necessary to **increase the** capacity of the electric **source is even** much **nearer than** would be at first imagined. **A single** really industrial installation **has** 120 baths, joined in **series,** worked from two **dynamo-** electric machines, also joined in series, **and we** believe that **this** proportion between **the** machines and **the baths should not be** exceeded to **remain in really economical conditions.**

It was Elkington, the inventor of industrial electro-silvering **and** gilding, **who** first refined copper by means **of alternating** rectified **currents.** The **patents** date **from 1866, and do** not embody anything important **as regards the composition** of baths or the use of electricity, but they are **very** explicit:— "The raw copper," it is **said,** "constitutes **the** plate which becomes dissolved **in** a suitable **solution;** pure copper is deposited on thin **surfaces connected to the negative** pole **as the** said plate is being dissolved. **It is advisable** to **use** a **series of vats, and to** work the entire system with **a** magneto-electric **machine.** The vats are charged with an almost saturated solution **of** sulphate of copper."

The electrolytic refining of **copper** has been practised for the **last** ten **years by the** Norddeutsche Affinerie at Hamburg, by Messrs. Oeschger & Mesdach **at Biache, by** M. Hilarion Roux

at Marseilles, by the Oker Foundry in Saxony, by the mines of
Mansfeld, by Messrs. Lyon-Allemand at Paris, by M. André
at Frankfort, &c., and by a few large English manufacturers,
notably Messrs. Elkington, and Elliott, of Selly Oak, near
Birmingham.

HAMBURG REFINING WORKS.—All refined coppers are not
equally good conductors of electricity ; their degree of purity,
and consequently of conductivity, is dependent on the nature
of the metal submitted to the electrolytic operation and on
the conduct of the operation.   The copper which has the best
reputation is that manufactured under the direction and on
the system of Dr. Wohlwill, by the Norddeutsche Affinerie
of Hamburg.   The current in this factory is obtained from
six Gramme machines, No. 1 type, and also from a much
more powerful machine specially constructed for M. Wohlwill
in 1873.

The admittance of the public to the works of the Norddeutsche Affinerie is prohibited, so that we have not been able
to obtain precise information as regards all the processes in use
at Hamburg ; we will describe further on the principal features
of the electric installation, but we can already assert that in
our opinion there is no manufacturing secret.   M. Wohlwill is
an intelligent chemist, well - informed in all the questions
relating to electricity ; he does not leave anything to the
unforeseen ; his baths are prepared with care, and always
maintained at the same temperature and the same degree of
concentration ; his machines revolve at regular speeds, and are
always in perfect working order ; his coppers are minutely
analysed both before and after the operation ; in short, installation, current, bath, manipulations, &c., everything contribute
towards a satisfactory result.   This is the sole reason of the
superiority of the coppers produced by the Norddeutsche
Affinerie.

The daily production of this manufactory reaches $2\frac{1}{2}$ tons
(2500 kilogrammes) of chemically pure copper.   The treated
copper often contains precious metals, which are found in the
bath after the operation.   The quantity of fine gold gathered
at Hamburg in 1880 reached as much as 1200 kilogrammes.

The following contain some details respecting the electrical installation proper :—

The first continuous-current machine executed for M. Wohlwill is illustrated (Fig. 32). It is provided with two collectors and four brushes. Each collector has twenty sections. The spirals of the bobbin are each composed of seven strips of copper 10 millimetres wide by 3 millimetres thick. Forty groups of copper ribbon correspond to the forty sections of the two collectors, so that each spiral is composed of two identical half spires juxtaposed and soldered at their extremities to a radiating piece which connects them to one of the sections of the double collector. The inducted ring is therefore composed of forty partial bobbins, of which twenty are connected to the right-hand side, and twenty to the left-hand side collector.

The total resistance of the induced bobbin is ·0004 ohm. When the two parts are joined in parallel this resistance is only ·0001 ohm.

The electromotive force, with a speed of 500 revolutions per minute, is equal to 8 volts for the coupling in series, and to 4 volts for the coupling in parallel.

The eight electro-magnets of this machine have iron cores 120 millimetres in diameter, and 410 millimetres in length. On these cores are wound thirty-two turns of sheet copper of the same width as the length of the electro, and a thickness of 1·1 millimetre. The resistance of the eight inductors in one single circuit is ·00142 ohm. When the electros are joined in two series, their resistance becomes ·00028 ohm. The total resistance of the machine is therefore ·00038 ohm in quantity, and ·00182 ohm in tension.

The total weight of copper used in the construction of the induced and the inducting coils is 735 kilogrammes. The machine is 1·50 metre long, ·75 metre wide, and 1 metre high. It weighs about 2500 kilogrammes. Its normal capacity is 3000 amperes when coupled in quantity, and 1500 amperes when coupled in tension. Its total production is therefore 12,000 watts per second.

The baths number forty, and are associated in two series of twenty. The surface of anodes immersed in each bath is about

Fig. 32.

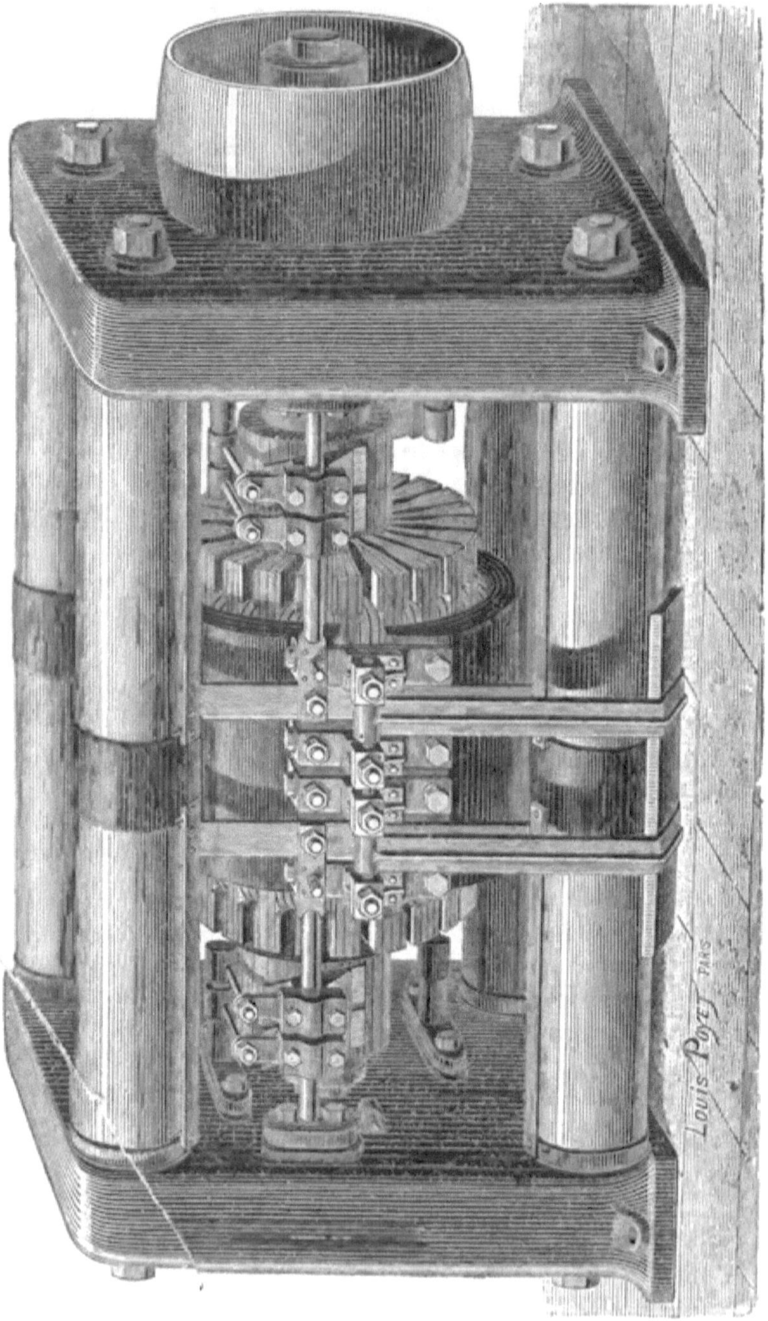

Gramme Machine for the Refining of Copper.

Louis Poyet PARIS

30 square metres, which **corresponds to** a total active surface of 1200 square metres.

The cathodes of refined copper **are about 1** millimetre in thickness.

The distance **between the** anodes and **the** cathodes is about 5 centimetres.

The quantity of copper deposited per **hour is 30·50** kilogrammes, and per day about 8000 kilogrammes. **The machine** has worked day and night for the past nine **years.** The expended motive power is **16** horse-power, which **corresponds** to 4,320,000 kilogrammetres per hour. Each kilogramme **of** copper thus treated therefore consumes 141,700 kilogrammetres (about one-half horse-power per hour).

The Norddeutsche **Affinerie possesses** besides, two other series **of baths,** which still **further economise the** motive power. These baths, **120 in number in each** series, are joined in tension. **Each** of them has 15 square metres of **surface** of anode, and a resistance **of ·00084** ohm. The **total resistance of the 120** baths is therefore **0·1** ohm.

The current **is** produced by **two Gramme machines No. 1,** joined in tension, and is **of a maximum capacity** of 300 amperes, with 27 volts of **total electromotive force at** 1500 revolutions per minute.

The quantity of copper refined is 900 kilogrammes per twenty-four hours.

The motive **power** expended **is 12 horse-power, which** corresponds **to** about 80,000 kilogrammetres per **kilogramme** of copper **refined.**

We **will** again refer **to this installation, which in** conception is the best in existence.

MESSRS. OESCHGER, MESDACH **& Co.'s** WORKS.—Messrs. Oeschger, Mesdach & Co. have installed in their factory at Binche-Saint-Waast (Pas de Calais), **a** Gramme machine identical with that which **had been built** for M. Wohlwill, and which **is illustrated p. 200.**

This machine supplies the current to twenty baths, and its daily production is **400** kilogrammes.

The vats are made **of** wood, nearly 3 inches thick, and lined with lead. They cost 350 francs (14*l*.) each, exclusive of the frames and conductors. Their length is 3 metres (10 feet), their width 80 centimetres (2 feet 8 inches), and their depth 1 metre (3 feet 4 inches). The two poles of the machine are connected with two brass frames made in one piece each. The negative frame is underneath the positive one and well isolated from it.

The baths are joined in tension. The positive frame of one being connected to the negative frame of the following bath, and so on, by means of a copper conductor so curved as to allow a free passage between the baths.

A galvanometer indicates the strength and direction of the current, and detects the polarising current. A cut-out automatically interrupts the operation when the counter-currents become too intense.

The bath is composed of a solution of sulphate of copper uniformly maintained at 19° Baumé. When the bath becomes too much charged with iron, it is purified by crystallisation.

All the baths communicate together at the bottom, so that their level is uniform.

The copper is deposited on copper cathodes suspended to the cross-bar of each negative frame. The cathodes are about 1 millimetre thick, and are simply folded at the top for hooking on the cross-bars of the frame.

The ores are treated as if it were only proposed to obtain ordinary merchantable copper; they are brought to a percentage of 95 per 100 in copper, and care is taken that this raw copper should not be too much charged with iron, sulphur, arsenic, antimony, &c. The copper, which in the bath is thicker at the top than at the bottom, is then melted and made into plates of about 1 centimetre average thickness.

In order to prevent certain parts, and notably those which are in contact with the top of the bath, wearing out too rapidly, they are coated with a non-conductive composition made of varnish and plumbic chromate.

These melted plates are suspended, by means of two copper hooks passing through two small holes made in them, to the cross-bars of the top frame, and serve as soluble anodes.

Each bath contains 88 anodes and 69 cathodes of equal total surfaces. The anodes are arranged in 22 rows of 4. They are 70 centimetres long, by 15 wide, and 1 (average) thickness. The cathodes are arranged in 23 rows of 3; they are 85 centimetres long, by 18 wide, and 1 millimetre uniform thickness. A calculation shows the double immersed surface of the anodes to be 1200 square metres. The distance between the anodes and the cathodes is 7 centimetres.

The copper is deposited on the cathode in thick layers, and sufficiently dense to be taken direct to the rolling-mill. The deposit is often detached, which is easily done, and melted again.

The silver falls to the bottom of the bath in a muddy form, together with the other impurities contained in the copper, and the pieces of anodes and cathodes which may become detached.

The baths are decanted by means of lead siphons, and the sediments are washed and sieved so as to separate the fragments of copper. The sediments are then melted with some litharge, or simply with a reductive flux. The matter thus obtained is cupelled with some argentiferous lead.

The liquors of the baths, which have been run into a lower decanting tank, are raised into a higher tank and used again. The raising of the liquors is effected by means of a small and very simple lead injector, which was, we think, invented by M. Thiollier.

Two workmen suffice for effecting the whole work, the manipulations being almost nil, and no accidental disarrangement in the bath installations or to the dynamo machine need be feared. In the course of five years only two stoppages of any importance took place; one of them was due to the renewal of the collectors of the Gramme machine; Messrs. Oeschger and Mesdach have since had a spare bobbin made, so as to avoid any possible repetition of a similar stoppage.

The director of the Biache works, M. Mascart, has carried out some experiments for determining the influence of soluble anodes in the refining of copper by electrolysis.

The following is a description of the results which he obtained :—

1st. One gramme of copper was dissolved in nitric acid and completely transformed into a sulphate. Three hours and fifteen minutes were required for depositing it with insoluble platinum anodes.

2nd. One gramme of silver, transformed into a sulphate and submitted to the action of the same current with an anode, also insoluble, was completely deposited in one hour.

3rd. Finally the same experiment took place with 1 gramme of copper and a soluble anode; 1·37 gramme was deposited in one hour, an amount which corresponds to 1 gramme in forty-three minutes.

It may be concluded from the foregoing, that in the conditions under which M. Mascart operated, the use of insoluble anodes requires, for effecting the same deposition, a current, or, what amounts to the same, an electromotive force four times greater than that of soluble anodes. We shall see that this proportion of 1 to 4 has nothing definite, and that the use of insoluble anodes always entails a considerable expenditure of current.

INSTALLATION AT FRANKFORT.—M. André, of Frankfort, has for some years carried out refining operations by means of Gramme machines. We do not believe that his processes, although ingenious, are used in practical installations.

His system was based on the electrolytic extraction of metals from mattes, speisses, &c., containing some nickel, cobalt, and copper, and on the introduction of a frame containing metal shots between the anode and the cathode. When a metal contained in an alloy is precipitated by electrolysis, the most positive metal contained in the frame precipitates a metal of the solution used.

The inventor also used conical revolving cathodes in order to avoid the polarisation.

M. André used sulphate of ammonium for the simultaneous precipitation of copper and nickel; dilute sulphuric acid for the separation of metals in old coins, and sodic chloride for the extraction of tin in old tin wares.

In the electrolysis of old coins the coins themselves were the anodes. Between the anodes and the cathodes a frame, covered on each side with cotton cloth and filled with copper shots, was placed. This frame divided the bath into two parts. Under the action of the current the silver and the copper dissolved in the dilute sulphuric acid; the copper of the frame precipitated the dissolved silver, and the copper only was deposited on the cathode, which was also of copper.

This process at first sight seemed somewhat complicated, and particularly when it was completed by conical revolving cathodes facilitating the gaseous escape; but M. André, whom we went to see at Frankfort, affirmed to us that he thus obtained economical results.

There unhappily are very few data respecting the processes in use in various factories for the electrolytic treatment of copper. The various manufacturers have devices of their own, their own compositions of baths, and their own methods for conducting the whole of the operation. The difficulty of protecting oneself against infringements prevents the manufacturers from taking out patents; on the other hand, they keep their means of production to themselves, and do not allow anybody to inspect their establishments. This certainly does not contribute to the rapid development of the electro-chemical science, but it is easy to understand that for a manufacturer or a commercial concern, private interests are stronger than public benefit.

Though not in a position to give details respecting all the large manufactories, we can give some information about the installations of M. Hilarion Roux, at Marseilles, and about the Elliott Metal Company's Works at Selly Oak, near Birmingham.

M. HILARION ROUX'S WORKS.—Mr. Mather has installed, at Marseilles, for M. Hilarion Roux, a small copper refining plant, of which the following are the particulars:—

Number of baths	40	
Total surface of anodes	900	square metres.
Surface of anode per bath	22·50	,,
Number of plates per bath	115	,,
Length of the plates	0·68	metre.
Width of the plates	0·15	,,
Thickness of the plates	0·01	,,

Weight of the plates	12 kilogrammes.
Height of immersion of each plate	0·58 metre.
Thickness of cathodes	0·0005 „
Distance between the anodes and the cathodes	0·05 „
Total weight of the copper treated	55 tons.
Gramme machine used	No. 1.
Number of revolutions per minute	850
Weight of copper refined per hour	10·40 kilogrammes.
Weight of copper refined per day	250 „
Weight of fuel consumed per day	240 „
Motive power required	5 H.-P.

The bath is composed of dilute sulphate of copper at 16° to 18°. The temperature is maintained constant at 25° C.

M. Roux's complete installation : baths, Gramme machine, and steam-engine, cost 1000*l.* The invested capital reaches 5000*l.* to 6000*l.*

THE ELLIOTT METAL COMPANY, LIMITED, WORKS.—This company, whose works are at Selly Oak, near Birmingham, refine 10 tons of copper per week. The current is produced by five Wilde dynamo-electric machines. Each machine feeds forty-eight baths joined in series.

The baths are 0·90 metres long by 0·90 metres wide and 1·20 metres high; they are lined with baked earth and a layer of asphalt. Each bath contains 16 anodes having 0·625 metre length by 0·175 metre width and 0·0125 metre thickness. The weight of one anode is 125 kilogrammes. There are 10 cathodes for each bath; they are 0·425 metre long by 0·400 metre wide and 0·0008 metre thick. The weight of one cathode is 1·30 kilogramme.

The total weight of copper contained in a bath is 205 kilogrammes, and the total weight of copper contained in a battery of 48 baths is 9840 kilogrammes.

The 'distance between the cathodes and the anodes is 0·0875 metre.

The anodes are immersed in the baths of 0·500 metre, so that the immersed surface is 0·5 metre × 0·175 metre = 0·0875 square metre for one side, and 0·1750 square metre for the two sides. The total immersed surface of anodes per bath is therefore 2·80 square metres.

There are five similar installations at Selly Oak Works. The weekly number of working hours is 156, which allows of a twelve-hours' rest on Sunday.

The production for forty-eight baths is: 13·5 kilogrammes per hour; 324 kilogrammes per twenty-four hours; 2268 kilogrammes per week.

Each bath therefore produces $\frac{13}{48}$ = 0·275 kilogramme per hour, which corresponds to a current intensity of 233 amperes.

The temperature in the bath-room is maintained uniform at 20° C. The degree of concentration of the baths is 16° Baumé. The anodes, having a thickness of 0·0125 metre, are replaced every five weeks.

The first Wilde machines, in use at Messrs. Elkington's and in a few large English factories, were, as we have explained in Chapter VI., composed of two superposed machines; one magnetic and of small dimensions, the other electro-magnetic and of large dimensions. The only duty of the small one was to excite the electro-magnets of the large one, the latter supplying the current to the refining bath. These apparatuses became, after a few hours' working, so much heated, that it was necessary to cool them by means of a circulation of water in the electro-magnets and the armatures. They entailed a considerable expenditure of work for a given quantity of electricity, and deteriorated somewhat rapidly. Notwithstanding these multiple inconveniences, it must be acknowledged that they rendered valuable service, and were superior to all the other known systems before the invention of the Gramme machines.

The Wilde machines used at Selly Oak are of an improved type (Fig. 33); but as in the previous types, they produce alternating currents, which have to be rectified with a commutator before sending them through the refining bath.

This new type consists of an armature provided with a series of bobbins revolving between the free extremities of a certain number of cylindrical electro-magnets arranged in a circle on each side of the said armature, and secured to the frame at their other extremities.

The armature bobbins are provided with iron cores, thus differing from the Siemens' alternating-current machines, in which there are no metallic cores. There are sixteen rows of bobbins and electro-magnets; two of the bobbins generate the exciting current, the fourteen others produce the current, which is used externally.

FIG. 33.

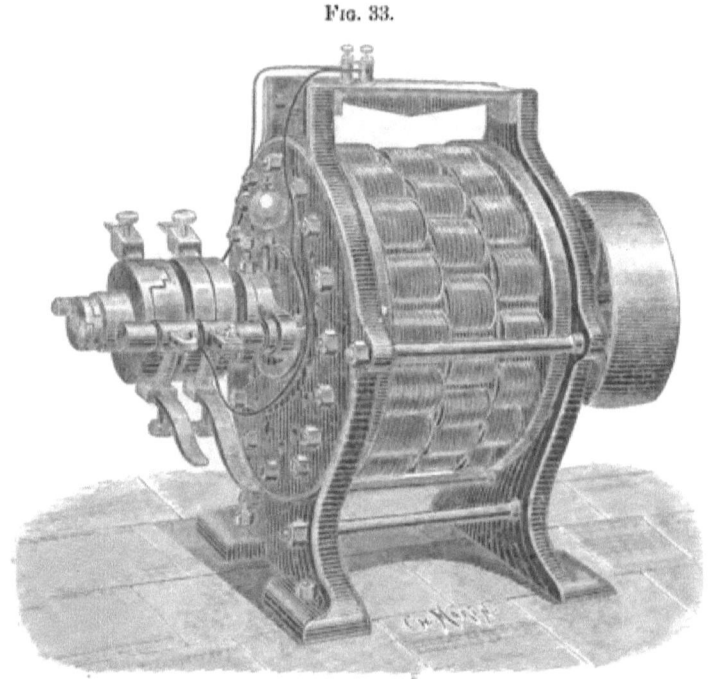

Wilde Machine for the Refining of Copper.

The machine is provided with two commutators; one for rectifying the currents traversing the inductor, the other for rectifying the principal current connected with the electrodes of the refining baths. These commutators are placed externally, and are easy of access for being turned, taken to pieces, or replaced in case of repairs or damage. This machine still heats much, but it does excellent service and is of great simplicity.

OKER'S FOUNDRY.—Mr. Siemens has constructed for the Oker's Foundry (Germany) some large dynamos, which are also used for the electrolysis of sulphates for copper refining purposes. There are three such machines, one having been at

work for the past four years, the others have been installed
more recently.

In order to give a low internal resistance and a great power
in amperes, the maker has surrounded the soft iron and the
bars of the electro-magnets with copper bars of rectangular
section, instead of winding them with wire, as in machines for
light or transmission of energy.

The induced bobbin is provided with a single series of bars,
with the Hefner-Alteneck system of winding. The corresponding
bars are joined together by means of large spiral bands situate
on the face of the bobbin opposed to the collector. On the
anterior face of the bobbin, on the collector side, the bars are
connected to the latter by means of strong copper angle pieces
(Fig. 34).

The inductor is also composed of a single layer of copper
bands coiled by series of seven on each branch, or twenty-eight
in all. The transversal section of these bars is 13 square
centimetres ; the junctions are bolted and soldered together in
order to secure good contacts.

The insulation of the bars is effected by means of asbestos,
which, owing to its non-conductivity, allows the machine to get
heated without danger. The machines used at Oker's reach to a
high temperature without any other inconvenience than that of
increasing their resistance, and consequently the motive power
necessitated for the precipitation of a given weight of copper.

Each of these machines feeds ten to twelve large vats, and
refines from 240 to 300 kilogrammes of copper per day (about
1 kilogramme per hour per vat). The power required for
the production of the current is from 10 to 12 horse-power,
corresponding therefore to 1 horse-power per vat.

The internal resistance of these machines is ·00075 ohm, the
electromotive force from 3 to 6 volts, the intensity from 810
to 1000 amperes. The total resistance of the baths and of the
conductors is about ·0055 ohm ; the relation between the
internal and external resistance of the machine therefore is
$$\frac{·00075}{·0055} = \frac{1}{7}.$$

We do not possess any further detailed particulars respect-

P

FIG. 34.

Siemens Machine for the Refining of Copper.

ing the installation. The foregoing demonstrates that the management of the Oker Foundry would not sink such an important capital as in the case of the works which we have described. The electrolysed copper appears to be difficult of refining, for the 'Electrotechnische Zeitschrift,' which first described the Oker Works, makes the following remarks:—

"These figures relate to cupric solutions containing not more than one-half per 100 of impurities. The polarisation produced in the vats is greater in proportion to the degree of impurity of the solution; and the work required to overcome this polarisation, which is at its maximum when liberation of gas, that is to say, decomposition of water, occurs, is also proportionally greater."

ECONOMICAL CONDITIONS OF ELECTRIC REFINING.—After having given a general idea of the processes employed in copper refining, and briefly described a few of the principal installations in which these processes are in use, we think it will be interesting to examine what are the conditions to be fulfilled in the establishment of a factory working economically, that is to say, in which copper is refined at the least possible cost.

In order to be in a position to formulate these conditions, it is necessary to review a few accessory questions, such as the maximum intensity of the current per metre of surface of electrodes, the resistance of the baths of sulphate of copper, the capital to be invested for a given production, &c., &c. The first of these questions is of the greatest importance, as it is the point of departure in all projects of installation for refining; we will therefore treat it with an exceptional development.

M. GRAMME'S EXPERIMENTS.—M. Gramme has carried out various experiments on sulphate of copper baths with soluble and insoluble anodes. The following is extracted from a note presented by him to the Academy of Sciences on the 29th of August, 1877.

When, after the creation of industrial dynamo machines, it was thought of using them for the operations of electrolysis, the customs of the factories where it was proposed to change the source of electricity had to be taken into consideration, and as in these works the baths were joined in parallel, it became necessary to construct machines of a very low internal resistance, so as to produce a great quantity of electricity with a comparatively low electromotive force.

The electromotive force increasing rapidly with the speed of the machines, M. Gramme was led to think that it would be possible to obtain much more considerable metallic deposits than were usually obtained without a proportional expenditure of motive power.

In order to as much as possible prevent the phenomena of polarisation, which would have modified the problem without facilitating its solution, M. Gramme took the particularly simple case of the electrolysis of sulphate of copper, using as anodes and cathodes copper plates of equal dimensions, and proposed to demonstrate that one bath offering to the current a given resistance and depositing a given weight of metal per hour, could always be replaced by two baths of the same nature having each half the resistance and depositing together twice as much metal in the same time; and this without any alteration in the expenditure of motive power, or in the electric current traversing the metallic solution.

First Series of Experiments.—The first experiments undertaken by M. Gramme were conducted with a variable number of baths all placed in parallel, that is to say, according to the old method. They proved that with one or thirty-six baths the deposition is the same with a constant intensity. This was a confirmation of Faraday's law, industrially realised with an excellent continuous-current dynamo-electric machine. Each bath contained only one anode and one cathode, having each 16 square decimetres of surface. The intensity of the current was 6·3 amperes. With one single bath the deposition was found to be at the rate of 7 grammes per hour; with twelve baths it was 7·1 grammes; and with thirty-six baths, that is to say, with a surface of anodes equal to 6 square metres, it was 7·1 grammes (about 2 decigrammes per bath).

Second Series of Experiments.—Table No. 1 gives the results of experiments conducted on baths joined in a chain, like the

TABLE No. 1. SOLUBLE ANODES.—BATHS IN TENSION.—INVARIABLE SURFACE OF ANODE FOR EACH BATH.

No. of the Experiments	Number	Electromotive Force (Baths)	Deviation of the Galvanometer	Total Weight of the Liquid in Action (kilos)	Temperature Initial	Temperature Final	Rise due to the Current	Work in Kgm. Total	Work Absorbed by the Frictions	Work Absorbed by the rising of the Temperature	Work Remainder	Deposition Total in 3 Hours	Per Hour	Per Bath and per Hour	Per Kilogrammetre of the Total Work, and Per Hour	Per Kilogrammetre of the Remainder
1	1	..	10·25	6·600	9·0	11·0	0·7	4·445	1·041	0·162	3·242	21·00	7·00	7·00	1·58	2·16
2	3	1·3	10·0	19·800	9·9	9·9	0·7	4·438	1·050	0·485	2·903	63·00	21·00	7·00	4·73	7·23
3	6	2·5	9·25	39·600	9·8	10·8	0·6	5·203	1·208	0·832	3·163	118·00	39·33	6·66	7·55	12·43
4	9	3·4	8·50	59·400	9·6	9·6	0·6	4·996	1·311	1·247	2·438	155·50	51·83	6·42	10·37	21·26
5	12	4·2	8·0	79·200	9·8	9·8	0·5	5·478	1·832	1·386	2·270	204·00	68·00	5·66	12·41	30·00
6	18	5·0	7·50	118·100	12·6	12·6	0·4	6·588	2·934	1·653	2·001	269·00	89·60	5·00	13·60	44·57
7	20	5·6	7·25	152·000	12·2	12·8	0·4	7·548	3·343	1·848	1·562	298·00	99·33	4·96	13·15	63·60
8	24	6·2	6·50	158·000	11·4	11·4	0·4	6·753	3·147	2·212	1·394	311·00	103·70	4·32	15·35	74·10
9	33	7·0	6·0	217·000	13·6	13·6	0·3	5·754	2·508	2·278	0·868	372·00	124·00	3·75	21·55	142·85
10	36	8·0	6·25	237·000	14·2	14·2	0·3	6·439	2·810	2·488	1·141	425·70	141·90	3·94	22·04	124·45
11	45	8·2	5·25	297·000	13·6	13·6	0·2	6·082	2·982	2·079	1·021	429·60	143·20	3·20	23·54	140·25
12	47	8·2	4·50	310·000	12·3	12·7	..	5·963	3·181	..	..	423·00	141·00	3·00	23·18	..
13	48	7·0	1·75	316·800	12·7	12·9	..	3·328	1·430	..	..	217·73	72·576	1·51	21·80	..

Observations.

1. The tension of the bath in experiment No. 1, and the rises of temperature in experiments 12 and 13, could not be determined.

2. The specific heat of the liquid was 0·89.

3. In the calculation of the rise of temperature due to the current, the ambient temperature of the laboratory has naturally been taken into account.

4. No. 13 experiment has been made with a machine having much more tension than the one used for the twelve first ones.

5. The resistance of the machine used for the twelve first experiments was equal to that of a copper wire of 1 millimetre diameter and 27·58 metres long.

TABLE No. 2. SOLUBLE ANODES.—BATHS IN TENSION.—VARIABLE SURFACE OF ANODES.

Nos. of the Experiments.	Number. N	Surface of each Bath in Square Centimetres.	Deviation of the Galvanometer.	Weight of the Liquid in action. kilos.	Temperature. Initial.	Final.	Rise due to the Current.	Work in Kilogrammetres. Total.	Absorbed by Friction.	Absorbed by the rising of the Temperature.	Remainder.	Deposition in Grammes. Total in One Hour.	Per Bath.	Per Kilogrammetre of the Total Work.	Per Kilogrammetre of the Remainder.	Observations.
1	3	8·26	7·5	19·8	13·04	13·9	0·3	3·397	1·722	0·624	1·051	15·75	5·25	4·63	15·00	1. The surfaces of the baths have been increased so as to obtain a constant deviation of the galvanometer. 2. The rise of temperature due to the current could not be determined in experiments Nos. 4 and 5.
2	5	16·52	7·5	33·0	12·5	12·9	0·2	3·452	1·765	0·693	0·994	29·00	5·80	8·43	29·17	
3	7	33·04	7·5	92·4	12·0	12·2	0·1	3·520	1·837	0·969	0·714	37·38	5·34	10·62	52·35	
4	9	49·56	7·5	178·2	13·0	13·1	‥	3·279	1·613	‥	‥	48·00	5·33	14·63	‥	
5	11	66·08	7·5	280·4	12·9	13·1	‥	3·449	1·788	‥	‥	61·60	5·60	17·85	‥	

TABLE No. 3. INSOLUBLE ANODES.

Nos. of the Experiments.	Number. N	Deviation of the Galvanometer.	Total Weight of the Liquid in action. kilos.	Temperature. Initial.	Final.	Rise due to the Current.	Work in Kilogrammetres. Total.	Absorbed by Friction.	Absorbed by the rising of the Temperature.	Remainder.	Deposition in Grammes. Total in One Hour.	Per Bath.	Per Kilogrammetre of the Total Work.	Per Kilogrammetre of the Remainder.	Observations.
1	1	11·00	6·6	10·3	11·5	0·9	7·615	2·459	0·627	5·529	7·07	7·07	0·921	1·26	1. In the first three experiments the baths were rejoined in tension; in the fourth one, they were all joined in quantity. 2. The secondary current was 70° (vertical galvanometer) with six baths in tension, 49° with three baths, and 16° with one. 3. In No. 4 Experiment the secondary current was 66°. 4. The duration of the first three experiments was one hour; that of the fourth was two hours.
2	3	8·25	19·8	10·2	11·1	0·5	8·504	4·302	1·044	3·158	16·33	5·33	1·885	5·06	
3	6	5·00	52·8	10·1	11·0	0·2	7·840	4·370	1·114	2·356	18·03	3·00	2·287	7·64	
4	12	11·00	79·2	11·6	11·9	0·2	7·615	2·459	0·835	4·321	7·50	62·0	0·981	1·66	

elements of a battery joined in tension; their number varied from one to forty-eight, but they all had electrodes of the same surface (16 square decimetres). The speed of the machine was increased as the number of baths grew larger, and the electro-motive force varied from 1 to 8 volts.

The figures of these tables establish that the deposition of the copper increased with the number of baths, and this not only as an absolute quantity, but also in a ratio to the work expended in the operation. The weight of copper per kilogram-metre varied from 1·58 gramme up to 23·18 grammes, and even up to 140 grammes if the loss of motive power which was found to occur, as we shall see further on, is taken into account; whereas in the first series of experiments the weight of the deposited copper did not exceed 1·96 gramme.

The practical conclusion of these experiments is self-evident; a great economy is afforded by joining the baths in series when dynamo-electric machines are used.

Third Series of Experiments.—M. Gramme purposed main-taining constant the intensity of the current in a series of comparative trials; this led him to increase the surface of the electrodes as well as the number of baths joined in series, so as to maintain constant the total resistance of the circuit.

Table No. 2 indicates that the quantity of copper deposited in a bath is much the same in all the experiments. The speed of the machine and the electromotive force of the cur-rent did not vary, and the work expended remained practically invariable.

These experiments are in perfect accordance with all the known theoretical notions except on one point; it will at once be observed that M. Gramme was driven to increase the sections of the liquid in a greater ratio than the number of baths joined in tension.

However this may be, various circuits are seen here, with uniform resistances and invariable electromotive force and in-tensity of current; it is therefore not surprising to see that in each part of these various circuits the quantity of deposited copper practically remains constant. But it will be observed that the total quantity of copper deposited in the complete circuit is proportional to the number of baths; from which it might be concluded that with a fixed expenditure of work it is possible by means of suitable arrangements to almost indefinitely increase the total deposition.

Fourth Series of Experiments.—In a last series of experiments M. Gramme has studied the effects due to the substitution—in sulphate of copper—of insoluble lead anodes for soluble copper anodes. As was to be expected, the polarisation was considerable, and the deposition of copper much less than previously, since the work of decomposition of the sulphate of copper was no more compensated by the formation of an equal quantity of sulphate, resulting from the attack of the anode by the acid.

All these experiments have been carried out with the utmost care, as M. Gramme explains in his report to the Academy of Sciences:—"I have," he said, "placed myself in conditions which I believe to be favourable for the measurement of the work expended in each experiment; the constancy of the work was nearly perfect during the three hours which each experiment lasted; I constantly verified it by galvanometric observations.

"At the termination of the experiment I opened the circuit and placed a Prony brake on one of the fly-wheels of the gas-engine, bringing it back to the speed at which it ran during the electrolytic operation, and I concluded from it what had been the expenditure of work.

"I could easily afterwards, by disconnecting the gas-engine from the dynamo, ascertain what proportion of the motive power was absorbed by the passive resistances of the latter. This quantity is given in the above three tables.

"I wished to go further, and ascertain the loss of work corresponding to the heating of the baths, and have arrived at the results by the following means:—

"In every experiment I took both the initial and final temperature of the baths; an inactive bath placed near at hand served as a means of comparison. The difference between the final temperatures of the active baths and of the inactive one represented the rise of temperature due to the current.

"Taking into account this difference, as also the quantity of the liquid operated upon, and the specific heat of the liquor, which I have found equal to 0·80, I obtained the number of calories supplied to the baths by the passage of a current. Multiplying then by the mechanical equivalent of heat, I obtained the quantity of work represented by this apparent heat.

"It will be understood that it is only the apparent and sensible heat of which I have thus been able to calculate the value; and that the results which I have obtained are inferior to the real figures.

"Deducting from the total work performed by the motor in each experiment the work corresponding to the friction of the electric machine and to the heating of the liquids, I obtained the quantity which I call remainder in the columns of my tables.

"In the experiments of the third series (Table No. 2), we have the irrefutable proof of the fact that the expenditure of work with soluble anodes in electrolysis can be taken as *nil*, for the deposition is seen to pass from 15 to 60 grammes without giving rise to an increase of work which could be measured. If the experiments of Table No. 1 show everywhere a remainder of work the use of which cannot be precisely determined, it must be observed that this remainder grows smaller as I realise some better conditions, and becomes as low as ·868 kilogramme-metre, and so less than one-sixth of the total work. It is explained by a calorific work in the other parts of the circuit.

M. BECQUEREL'S EXPERIMENTS.—M. Ed. Becquerel carried out, more than twenty years ago, some very interesting researches on the electrolysis of the sulphate of copper, with a view of determining if the quantities deposited on the cathodes exactly corresponded to the quantities dissolved by the anodes. These experiments[*] have been conducted simultaneously, with the solutions of sulphate of copper sold in the trade, of copper acidulated by means of sulphuric acid diluted to $\frac{1}{20}$, of sulphate of copper crystallised several times, and of sulphate digested in a cold state with carbonate of copper so as to render it neutral. M. Becquerel used a number of Bunsen cells varying from one to ten; the weights of copper deposited varied from 1 to 15 grammes.

[*] 'Eléments d'electro-chimie,' by M. Becquerel, edition 1864, p. 220.

He obtained the following results:—

1st. With the saturated solution of sulphate acidulated to $\frac{1}{20}$, the loss in weight of the copper electrode at the positive pole is always greater than the increase in weight of the cathode. This difference has varied between $\frac{2}{100}$ and $\frac{5}{100}$ of the weight of the deposition (or an average of 0·033). Care was taken after withdrawing the electrodes from the solution to dry them *in vacuo*, so as to prevent the alteration of the copper.

2nd. With the solution of sulphate of copper neutralised in a cold state by the carbonate, the loss at the positive pole has been sometimes greater and sometimes smaller than the gain at the negative pole, but the differences did not amount to $\frac{1}{100}$ part of the weight of the deposition. On an average, the loss of weight was smaller than the weight of the deposition by a few thousandths only. The deposition in the two first solutions (acid and ordinary) was very close and cohesive, and could be taken off in bands; whereas in the two last (pure and neutral) the copper was crystallised and more cohesive.

But the experiments, especially with the fourth solution, are not satisfactory in this respect, that some blackish scales produced by the formation of subsalts drop from the positive pole; which does not occur in the first ones, owing to the acidity of the solution. In the same neutral solution the negative cupreous deposit is pink-tinted, and probably contains some protoxide, for it loses a part of its weight if it is heated in hydrogen.

These results illustrate the influence of the liquids and of the electrodes on the effects produced and on the cohesion of the deposition. If, however, the results obtained with the three first solutions are considered, it may be admitted that the differences in the loss of weight of the soluble anode is principally due to the acidity of the solution, since the sulphate of copper sold in the trade is always acid, and with the pure sulphate this difference is very small.

If we describe in detail M. Becquerel's experiments, it is not only to warn the refiners against a loss of work amounting to from 3 to 4 per cent., which loss can easily be avoided by using sulphates as pure as possible, but more particularly to call their attention to the necessity of preventing the variations of composition of their baths. With solutions too rich in copper, the resistance, and consequently the motive power, would be reduced, but the deposition of the metal would be less perfect, and its tenacity smaller.

In another part of his treatise on electro-chemistry, M. Becquerel describes a curious fact, which we have also noticed, and which we think is worth quoting.

" When sulphate of copper is prepared on a large scale and stored in wooden vessels, a deposition of metallic copper can be observed to take place, after a certain time, at the extremities of some of the staves. The organic matter intervenes to operate the reduction. The deposition of copper gradually increases until large masses of adhesive copper are formed. In this reduction the dioxide sulphate is changed into protoxide sulphate. But there remains to be explained why the deposition occurs in certain places and why all the parts offer the same cohesion. Let us admit, as everything seems to prove it, that the first particles of copper deposited have been so deposited by the reaction which we have mentioned. These particles, which are in contact with the carbonaceous matters of the decomposed wood, constitute a voltaic couple; the copper is the negative pole, and the carbonaceous matters on which the sulphuric acid reacts is the positive one: consequently the copper must be deposited on the copper already precipitated."

THICKNESS OF THE DEPOSITION OBTAINED IN A GIVEN TIME. —We will attempt to determine the cost of the electric refining of copper in various conditions of installations; to that effect we must first calculate the possible thickness of deposition per hour, then the expenditure of power required by the electrolysis, and lastly the capital invested in the manufacture.

In his work on the theory and the applications of electricity, Mr. John T. Sprague describes his personal experiments on the deposition of copper effected in a bath composed of 3 parts of a saturated solution of sulphate of copper and 10 parts of sulphuric acid diluted in 10 times its volume of water.

The operation was not stopped before the thickness of the deposition reached 52 centigrammes for 6·45 square centi-

metres, which corresponds to 8 centigrammes per square centimetre. The thickness was 0·09 millimetre, taking the density
of the precipitated copper at 8·89.

Mr. Sprague used, as a generator of electricity, a Daniell
cell, and he varied the resistances so as to obtain a thickness of
0·09 millimetre in thirty hours at a maximum, or in forty-five
minutes at a minimum.

This is the result of the experiment :—

	hrs.	min.		mm.		
No. 1. In	30	0	thickness of	.003	per hour.	Excellent deposit.
2.	15	0	,,	·006	,,	Copper of great tenacity.
3.	5	0	,,	·018	,,	Very good deposit.
4.	2	30	,,	·036	,,	Good deposit.
5.	1	15	,,	·072	,,	Sandy at the edges.
6.	0	45	,,	·122	,,	Bad deposit.

The first four deposits were practically equally regular and
homogeneous, but if the deposition is accelerated so as to exceed
a thickness of ·036 millimetre per hour it becomes defective.
Mr. Sprague consequently recommends not to exceed that limit,
which corresponds to about 1 ampere per 33 square centimetres,
or 300 amperes per square metre of surface of anode.

This electrolytic regimen cannot in practice be attained, for
the manufacturers do not proceed with the same care as Mr.
Sprague ; their baths often are only slightly homogeneous, the
anodes and cathodes irregularly disposed, and the deposit
becomes coarse and without adhesion much before that limit.
We only recommend it in one case, that is, in the production of
electrotypes for the reproduction of wood engravings, when a
prompt delivery is required ; and we would still insist that
the first coating should be slowly deposited, and that the rapid
deposition should only apply to the thickening of the coating. To obtain a good result 1 ampere per square decimetre,
or 100 amperes per square metre of surface of cathode should
not be exceeded when the anodes are of chemically pure
copper.

In the experiments which we have described, M. Gramme,
placing himself at an entirely different point of view to that of
Mr. Sprague, and wishing to absolutely avoid the irregularities
in the quality of the depositions, so as not to arrive at any

erroneous conclusion, sometimes used a current of 6·3 amperes per 1600 square centimetres, which corresponds to 1 ampere per 254 square centimetres, or 30 amperes per square metre; and sometimes the same current of 6·3 amperes for 6 square metres, that is to say approximately 1 ampere per square metre. Under these conditions, the deposition was naturally regular, fine and cohesive, and the thickness obtained in one hour varied from ·004 millimetre to ·0001 millimetre.

At Hamburg, M. Wohlwill, having always had in view the object of economising the motive power, has also been led to operate with large surfaces of cathodes, obtaining, therefore, small thicknesses of copper depositions per hour. In his first installation the large Gramme machine deposits 30·50 kilogrammes per hour on 1200 square metres of surface, that is to say 25·4 grammes per square metre. The thickness of the copper precipitated on the cathodes is therefore ·003 millimetre per hour. In the two installations which followed, this result was still very much reduced since the deposit per hour did not exceed 37·50 kilogrammes of copper on 3600 square metres of cathodes, which corresponds to an approximate thickness of ·001 millimetre in the same time.

At Messrs. Oeschger and Mesdach's, the production is 700 kilogrammes of copper in 24 hours, or 29 kilogrammes per hour, with 1200 square metres of surface of cathodes. The thickness of the deposition is approximately ·003 millimetre per hour.

At Marseilles, M. Hilarion Roux precipitates 10·4 kilogrammes of copper per hour on 900 square metres of cathodes, or 11·44 grammes per square metre. The thickness of the deposition is therefore ·001144 millimetre per hour.

The Elliott Metal Company at Birmingham have five installations which precipitate each 13 kilogrammes of copper on 144 square metres of cathodes, or 90 grammes per square metre. The thickness of the deposition is, in the five installations, ·01 millimetre per hour.

In M. Michel's factory in Paris, electrotypes ·1 millimetre thick are obtained in five hours, which correspond to ·02 millimetre per hour.

It would be useless to multiply the examples, for in all those which we have given the deposits vary between ·0001 and ·1220 millimetre per hour and comprise all the possible applications.

In conclusion, we have calculated the thicknesses of the depositions which could be obtained per week of 156 hours with the ten cases which we have described, and the following are the results which we arrived at :—

THICKNESSES OF COPPER DEPOSITED IN 156 WORKING HOURS.

Experiments and Applications.	Thicknesses in Millimetres.
Experiments of M. Gramme	·0156 to ·624
„ Mr. Sprague	Good deposit : ·468 to 5·616
„ „	Bad deposit : 11 to 19
Works of Messrs. Oeschger and Mesdach ..	·468
„ The Norddeutsche Affinerie	·468
„ „ „	·156
„ M. Hilarion Roux	·178
„ The Elliott Metal Co.	1·56
„ M. Michel, Paris	3·12
Maximum regimen, chemically pure anodes..	1·68

It has been established from chemical analyses, that the purity of the copper electrolytically obtained, depends upon the distance between the anodes and their cathodes, upon the continuity of the operation, the composition of the baths and the nature of the foreign substances contained in raw copper ; but that it is not altered, all things being equal, when not more than 2 millimetres thickness per week are deposited with a normal regimen. In this respect the works of Hamburg, Marseilles, and Birmingham show no real superiority over each other.

SPECIFIC RESISTANCE OF THE BATHS.—The electromotive force required for the electrical refining of metals, is, in a great proportion, absorbed by the resistance of the baths, which resistance depends on the degree of concentration of the baths, their temperature and the distance of their electrodes.

Numerous experiments have demonstrated that the best solution to employ in the electro'ytical refining of copper is one at 16° to 18° Baumé.

This solution corresponds to 18·267 per 100 in weight of $SO_4Cu + 5HO$.

Its density at 16° is 1·1247.

The equivalent of $CuSO_4$, 5HO = 124·7.

The equivalent of $CuSO_4$, = 79·7.

The relation between the equivalent of the solution and of the sulphate of copper is

$$\frac{CuSO_4}{CuSO_4 + 5HO} = \frac{79·7}{124·7} = 0·68,$$

from which, for a density of 1·1247, we have

$$18·267 \times 0·68 = 12·42 \text{ of } CuSO_4.$$

The density of the bath being known, the distance, according to the temperature, can easily be determined by means of the tables which we have published in the first portion of the work. At 20° C. a solution of 12·42 per 100 of $CuSO_4$ has a specific resistance of about 32 ohms.

So that 1 cubic centimetre of the bath, considered between its two opposite faces, has a resistance of 32 ohms.

This resistance is 10,000 times smaller per square metre or ·0032 ohm for a thickness of 1 centimetre. With 2, 3, 4, 5, . . . centimetres, the resistance is naturally 2, 3, 4, 5 times greater.

We have ascertained from direct experiments that the baths of sulphate of copper used in refining were much less resistant than shown in the previous calculations. This result is due to the facts that the heat developed in the electrolyte owing to the passage of the current, increases its conductivity, and that the degree of acidity which increases after a few hours working diminishes the resistance of the bath in a comparatively large proportion.

In estimating the resistance, the layer of liquid situate between the anodes and the cathodes should not be exclusively considered, for the total conductivity depends not only on that layer, but also on the underneath and lateral expansions of the said layer, which expansions are often considerable compared to the surface of the electrodes.

The following are a few of the results which we have obtained in operating on sulphates of copper of the trade :—

1st. Temperature 20° C. ; concentration 10° Baumé; specific resistance 25 ohms.

2nd. Temperature 20° C. ; concentration 18° Baumé; specific resistance 20 ohms.

3rd. Temperature 25° C. ; concentration 18° Baumé; additional acid 1 per 100 ; specific resistance 15 ohms.

The foregoing show that it is easy to establish some refining baths having a specific resistance of 20 ohms, and we believe that the solutions in use at Hamburg and Birmingham have a specific resistance lower than 20 ohms; we will, however, take this figure as the basis of our calculations, because we prefer the use of resisting baths to that of acid ones, as regards the purity of the copper obtained.

The distance between the anodes and the cathodes varies from 2 to 10 centimetres. In refining work the minimum distance is scarcely less than 5 centimetres, in order to facilitate the manipulations and prevent any contact between the positive and negative plates. We have, however, seen installations in good working order with a distance of only 4 centimetres, but this necessitated an extra quantity of labour for fixing the anodes and the cathodes on the cross bars of the baths. The resistance of refining baths varies between ·01 and ·03 ohm per square metre of surface of anode.

It will thus be seen that the greater the surfaces of anodes for a given hourly deposition, the smaller is the resistance which the current has to overcome in order to effect the precipitation of copper ; consequently the power required for driving the dynamo is smaller, or, what amounts to the same, the smaller the thickness of deposition per hour, the smaller the motive power required. If the operations took place on infinitely large surfaces, large quantities of copper would be deposited with an infinitely small power. This is an important fact which M. Gramme had the honour of being the first to put in evidence and upon which we could not too much insist.

Let us now examine what is the total resistance of the baths in a few of the applications which we have described.

At Marseilles, the surface of anodes is 22 square metres per bath; the electrodes are 5 centimetres apart, and there are forty baths. The resistance for each bath is—

$$R = \frac{0 \cdot 0020 \times 5}{22} = 0 \cdot 00046 \text{ ohm},$$

and for the **40 baths 0·0184 ohm.**

At **Hamburg** (we only take into consideration the last two installations, as they constitute a progress on the first one), we will determine the distance between the anodes and the cathodes, knowing the resistance of one bath.

This resistance is equal to that of a copper wire 5 millimetres diameter and 1 metre long, that is to say **0·00084 ohm** for one bath, and 0·1 ohm for 120 baths.

The surface of anodes being 15 square metres per bath, **the** resistance per square metre is 0·00084 ohm × 15 = **0·0126** ohm, **the** distance between **the anodes** and the cathodes **is** therefore :

$$E = \frac{0 \cdot 0126}{0 \cdot 002} = 6 \cdot 3 \text{ centimetres.}$$

At Birmingham the average distance between the electrodes is 6 centimetres, and the **surface** of **anodes per bath** $2^{mq} \cdot 80$; the resistance of one bath is therefore :

$$R = \frac{0 \cdot 0020 \times 6}{2 \cdot 80} = 0 \cdot 00428 \text{ ohm per bath,}$$

and 0·20544 ohm for the **48 baths.**

The electrical work performed **by** the passage of the current through these various resistances is easy of calculation when **the** intensity of the said current is known. **For** instance we have seen that at Marseilles 10·40 kilogrammes of copper were precipitated per hour in **40** baths, that is **to say,** 260 grammes **per** bath per hour ; **at** Hamburg the deposition **per** bath per **hour** is 312 grammes, and at Birmingham 270 grammes.

We also know that each ampere liberates **1·18 gramme of** copper per hour ; the intensity of the **current therefore is** :

$$C = \frac{260}{1 \cdot 18} = 220 \text{ amperes, at Marseilles.}$$

$$C_1 = \frac{312}{1 \cdot 18} = 265 \text{ amperes, at Hamburg.}$$

$$C_2 = \frac{270}{1 \cdot 18} = 230 \text{ amperes, at Birmingham.}$$

The work absorbed by a resistance R which is traversed by
a current of an intensity C, is given by the formula :—

$$W = \frac{C^2 R}{g}.$$

If we apply this formula to the three applications which
are actually being analysed we find :

Marseilles $W = \dfrac{\overline{220^2} \times 0 \cdot 0184}{9 \cdot 81} = \begin{cases} 89 \cdot 06 \text{ kgm.} = \text{about} \\ 1 \cdot 2 \text{ h.p.} \end{cases}$

Hamburg $W_1 = \dfrac{\overline{265^2} \times 0 \cdot 1}{9 \cdot 81} = 705 \text{ kgm.} = 9 \cdot 5 \text{ h.p.}$

Birmingham $W_2 = \dfrac{\overline{230^2} \times 0 \cdot 2054}{9 \cdot 81} = 1107 \cdot 6 \text{ kgm.} = 14 \cdot 7 \text{ h.p.}$

The work absorbed by the metallic resistances is very small,
and can be reckoned at 1 per 100 of the work absorbed by the
bath.

The polarisation of the electrodes, which particularly depends
on the composition of the raw coppers, consumes a mechanical
energy varying between 5 and 10 per 100 of the total work.

Lastly, the efficiency of the machines attains 90 per 100
with a good utilisation, and decreases to 70 per 100 when the
machine heats and is not kept in good working order.

Knowing the elements of the three preceding installations,
we have been able to establish with sufficient correctness the
following table, which can be used as a basis to a project :—

Designation of the Factories.	Weight in Kilogrammes of the Copper Refined.		Work in Kilogrammetres.					Total Work in Horse-power.
	Per Hour.	Per Horse-power.	Per Kilogramme of Copper.	For traversing the Baths.	For the Machines.	For the Polarisation and other unde-termined causes.	Total.	
Norddeutsche Affi-nerie, Hamburg	37·50	3·12	24	705·00	105	91·00	910	12
Hilarion Roux, Marseilles ..	10·40	2·65	29	89·06	82	128·94	300	4
Elliott's Metal Co., Birmingham ..	13·00	0·67	113	1107·C0	200	152·40	1400	19

There results from the foregoing calculations that by increasing the number of baths and the surface of the electrodes of each bath, the saving of three-quarters of the motive power expenditure can be practically effected. When the fuel is dear it is absolutely necessary to work in this fashion, otherwise the cost of electrolytical refining would render such an operation industrially impracticable.

CAPITAL ABSORBED.—In order to determine the amount of capital absorbed in an installation of copper refining, we will still take as examples the works of Marseilles, Hamburg, and Birmingham ; not that we know exactly what these installations did cost, as the expenditure must naturally have been considerable, owing to the fact that at first the work must necessarily have been of an experimental kind. If we consider these factories here it is only because of their importance, and of the system adopted by them for the grouping of their baths, and we will calculate what would be the cost of similar installations.

One vat of one cubic metre, for refining purposes, cost, empty, about 130 francs, and with the solution of sulphate of copper 200 francs ; this total price is doubled for a 3 cubic metres vat.

The metallic conductors generally cost 2·50 francs per kilogramme, 100 kilogrammes per bath of 1 cubic metre should be reckoned upon, which corresponds to an expenditure of 250 francs. This expenditure is also doubled in the case of a 3 cubic metres bath.

A good steam engine costs, for 4 to 5 horse-power, about 4500 francs ; for 10 to 15 horse-power, about 9500 francs, and for 20 horse-power about 12,000 francs.

The dynamo-electric machines used for electrolytic purposes, can be valued on a basis of 500 francs per electrical horse-power, which corresponds to 7500 volt-amperes. /7 50

With these particulars we can already condense the costs of first establishment of these three factories.

The quantity of copper to be refined, which plays the principal part in the absorbed capital, is inversely proportional to the rapidity of the deposition. At Hamburg, in one installation only, there is about 1200 square metres of cathodes and

1200 square metres of anodes, including the parts which are outside the baths. The thickness of the anodes is 10 millimetres, that of the cathodes 1 millimetre. The total weight of the copper under treatment is 125 tons.

COST OF FIRST ESTABLISHMENT.

Factories taken as Types.	Number of Baths for one Installation.	Capacity of Baths.	Motive Power Necessary.	Expenditure.					Per Ton of Copper Refined per Year.
				For Baths.	For Conductors.	For Motors.	For Dynamos.	Total.	
	c.m.	c.m.	h.p.	frs.	frs.	frs.	frs.	frs.	frs.
Hilarion Roux, Marseilles	40	3	..	16,000	20,000	4,500	2,500	42,000	466
Norddeutsche Affinerie, Hamburg	120	2	12	36,000	48,000	9,500	5,000	98,500	310
Elliott's Metal Co., Birmingham ..	48	1	20	9,600	12,000	12,000	9,500	43,100	414

The market price of raw copper varies; this may be prejudicial to manufacturers who slowly treat large quantities of it at a time in the case of a persistent decline in the price. As we are only comparing here a few systems of installation, we may, to fix the ideas, assume that the price is 1600 francs (64*l*.) per ton (this is the average price for the last two years).

The value of the copper under treatment, in one of the installations of the Norddeutsche Affinerie at Hamburg, therefore represents the enormous sum of 200,000 francs (8000*l*.). Assuming an annual production of 330 tons, we see that for each ton refined there must constantly be a value of 606 francs (24*l*. 5*s*.) of copper in the bath.

At Marseilles 55 tons of copper are immobilised, which represents a capital of 73,000 francs (2920*l*.) for an annual production of 80 tons or 910 francs per ton of refined copper.

At Birmingham there is only 9840 kilogrammes (about 10 tons) of copper under treatment, of the value of 15,744 francs (630*l*.), for an annual production of 104 tons, which reduces this part of the absorbed capital to 151 francs per ton of copper refined.

The ground space occupied is also an item in the expenses of establishment, an item which very much varies according to

countries, but may be estimated at a minimum of 100 francs
per square metre on account of the buildings which have to be
erected on these spaces.

The installation of the 120 baths at Hamburg, the passages
for the service, the engine and electric machine rooms, &c.,
occupy a space of 660 square metres, which calculated upon
the basis of 100 francs per square metre, corresponds to an expenditure of 66,000 francs for 330 tons per year, or to 200 francs
per ton.

At Marseilles, with 40 baths, we must admit a space of 300
square metres for the whole of the services. That surface can
be valued at 30,000 francs for an annual production of 80 tons,
or at 375 francs per ton.

At Birmingham, 160 square metres may be sufficient for a
production of 104 tons, which corresponds to a total expense of
16,000 francs, or to 154 francs per ton of copper.

By adding together these various expenses, we obtain the
total outlay per ton of refined copper per year.

	First Establishment.	Metal in the Vats.	Ground Space.	Total Outlay.
	frs.	frs.	frs.	frs.
Marseilles..	466	910	200	1576
Hamburg	310	606	375	1291
Birmingham	414	151	154	719

COST PRICE OF THE REFINING OF COPPER.—The expenditures incurred in the electrolytic treatment of copper are as
follows:

1st. The interest of the engaged capital.

2nd. The fuel and accessories required for the motor.

3rd. The labour.

4th. The maintenance and redemption of the material.

5th. The general expenditure.

A rate of 5 per 100 on the engaged capital can be admitted,
although the refining works generally belong to dealers in
metal, who have almost always a large stock of copper to guard
against the fluctuations of the market, and it is consequently, if
not absolutely immaterial to them—at least not onerous—
whether this copper is in the baths or in stock. However, in the

case of a rapid and persistent decline in the prices the large stocks may involve some important losses; that is why an interest of 5 per 100 must be maintained every year in view of an abnormal situation which rarely occurs.

We can estimate the cost of fuel at 20 francs per ton, although that price would be too high in the case of Birmingham, sufficiently approximate for Hamburg, and quite insufficient for Marseilles. However, as we are only making a comparison we will maintain a uniform price; it will always be easy afterwards to recalculate the cost, taking as a basis the actual cost of fuel in the locality considered.

An engine of 4 to 5 horse-power consumes 20 kilogrammes of fuel per hour; the wages of the driver being estimated at 60 centimes per hour, and the accessory expenses of waste, grease, &c., at 40 centimes per hour, the total hourly cost of the motive power is therefore 1·60 franc approximately.

A 20 horse-power engine consumes 50 kilogrammes of fuel per hour; the driver's wages being 70 centimes per hour, and the accessory expenses 60 centimes. Total, 2·30 francs per hour.

The cost of maintenance and the wear and tear of the apparatus in use represent a minimum of 10 per 100 of the purchase price; those of the buildings 5 per cent. The electric conductors, which convey the current in the baths, do not alter in price. Although the market price may fluctuate we cannot introduce such variations into our calculations.

The labour, in a factory of 40 baths, amounts to 75 centimes per hour, or 18 francs per day; it is double this amount in a factory of 120 baths.

The general expenditure can be estimated at 100 per 100 of the cost of labour in large installations of 200 to 300 baths for example, and at 150 per 100 of the cost of labour in installations of only 40 to 50 baths.

The preceding figures will enable us to establish a comparative table of all the expenses necessitated by the electrolytical refining of copper.

We cannot too much insist that these figures are not absolute; they however are sufficiently approximative for serving as a basis to a project. They convey an exact idea of the ele-

COST PRICE OF REFINING COPPER BY ELECTRICITY.

Factories taken as Examples.	Expenditure per Ton of Refined Copper.					
	Interest on Capital.	Motive Power.	Maintenance.	Labour.	General Expenditure.	Total.
	frs.	frs.	frs.	frs.	frs.	frs.
Hilarion Roux, Marseilles..	78·80	112·00	18	72·00	108·00	388·80
Norddeutsche Affinerie, Hamburg .. ..}	64·65	39·50	12	40·00	40·00	196·05
Elliott's Metal Co., Birmingham .. ..}	35·95	180·06	30	57·75	57·75	361·45

ments which constitute **the cost of the** electrolytical refining of **copper, and their true interest** consists in the comparison **which they allow of** being established between various factory **installations.**

The cost of fuel in Birmingham **is much** lower than **that** which **we have** taken as a basis; **but** taking **it** at 6 francs per ton at the **works we** find that the motive power still costs 1·20 franc per hour, **or** 125 francs per ton of copper. If we leave all the other figures unaltered we obtain **a total of** 306·45 francs, that is to say, **a much** greater expenditure **than at** the Hamburg factory. The **interest on** the capital engaged represents a small proportion **only of the cost price,** whereas **at** Hamburg it con-stitutes the main expenditure.

As it was **easy to foresee,** two factories, those **of Hamburg and** Marseilles, established **with** the same elements **and** on the same lines, give essentially different results in **their** working, owing to their **respective magnitude.** At Hamburg, where the operations **are conducted on a large scale,** the cost price of refining is **about** 200 francs, **whereas at Marseilles,** where the works are not **of** much importance, **this cost is nearly doubled.** The arrangement of 120 baths in tension and the considerable surface of anodes is much to be preferred to that **of** 48 baths of small surface, notwithstanding the enormous capital sunk in the first case.

If **water, instead** of steam power, were used, it would still be necessary, for economically refining **the** copper, **to** adopt the disposition in use **at** Hamburg.

The work obtained from a water-wheel or from a turbine is

not as cheap as some people fancy. The lubricating and main-tenance of the motor, the repairs to the valves, the over-fall, the canals of derivation and the mill race are the cause of an annual expenditure which sometimes proves very important. We know of some factories where the hydraulic power costs as much as that supplied by a steam engine with fuel at 10 and even 15 francs per ton. The flow of water is generally irregular, and it often happens that the adjunction of a steam engine is necessitated in order to prevent a partial stoppage of an esta-blishment situate on a watercourse. This naturally increases the cost of first establishment, and consequently the total ex-penditure per ton.

By combining together all the best possible arrangements—hydraulic power, large number of baths in tension, minimum of copper under treatment, &c.—we believe a manufacture could be established in which the cost of refining copper would not ex-ceed 150 francs per ton (6l.); but it would be difficult, if not impossible, to obtain it at a lower price, and especially if nothing was neglected in order to obtain very pure copper of a high electric conductivity.

In conclusion, the refining of raw copper by means of elec-trolysis necessitates much space, considerable sums of money, and is a very expensive process. It is only just to add that raw coppers generally contain a small proportion of gold and silver, and that the cost of refining is often compensated, and even more, by the value of the sub-products; * on the other hand, the copper is much purer than that obtained from metal-lurgical treatment, and its high conductivity renders it a much more valuable metal.

REFINING OF LEAD.

The following is, we believe, the only process applied to the industrial refining of lead.

The inventor is Mr. Keith. The Company working his pro-

---

* The sale price of raw copper to refining works has naturally been affected by the knowledge that they contain a proportion of precious metals; but in general the industry benefits by this treatment, owing to which excellent coppers are offered in the market at comparatively moderate prices.

cess is the Electro-Metal Refining Co., of New York; its capital
is 2,500,000 francs (100,000*l.*).

The lead to be refined is melted in plates, and these are
suspended to metallic cross-bars connected with the positive
pole of a magneto-electric machine; between these plates are
placed cathodes of pure lead connected with the negative pole.

The bath is a solution of plumbic sulphate in sodic acetate.
The metals acting as positive towards lead, such as iron and
zinc, remain almost entirely in the solution, and are precipitated in a state of oxide only, easy to separate when the
metal is being melted afresh.

The plumbic sulphate is decomposed by the action of the
current, the lead going to the cathode and the acid on the
anode, where it dissolves the lead, the iron, and the zinc of the
metal to be refined. The gold, the silver, and the antimony
remain on the anode, where they are gathered in muslin
bags.

The following are, according to Mr. Keith's experiments, the
results obtained when operating on a few tons of base bullion.

With 48 wooden vats, containing each 50 plates of base
bullion weighing 16 kilogrammes each, and 12 horse-power, the
production reaches 10 tons per 24 hours.

The plates are of the following dimensions:—

Length	1·22 metre.
Width	0·38 ,,
Thickness	0·003 ,,

This is the composition of the lead before and after refining:—

	Before refining.	After refining.
Lead	96·36 per 100	99·9 per 100
Silver	0·5544 ,,	0·000068 per 100
Copper..	0·315 ,,	0·0
Antimony	1·070 ,,	traces
Arsenic	1·22 ,,	traces
Zinc, iron, &c. ..	0·4886 ,,	0·0

The sediments gathered in the muslin bags are dried
and melted in a crucible, with sodic nitrate and borax. The
silver remains in a metallic state; the arsenic and antimony

form a scoria which is treated by hot water; the sodic arse-
niate dissolves, and it is crystallised; the sodic antimoniate
is reduced by the fire; the iron and copper which it may contain
are not reducible.

The actual treatment of the base bullion by the dry method
costs about 30 francs per ton; Professor Barker, of Philadelphia,
estimates that by using the Keith process the cost would not
exceed 10 francs. The saving would, therefore, amount to 20
francs per ton. Dr. Hampe, speaking on this process, records
the fact that the lead deposited upon the cathode is not
absolutely pure, and contains in particular a certain proportion
of bismuth. The following are the results which he has obtained when electrolysing 6 litres of plumbic acetate containing
77·92 grammes of lead per litre, and acidulated with 4 per
cent. of acetic acid.

The surfaces of the electrodes were 13,000 square millimetres.

	Raw Metal.	Deposited Lead.	Argentiferous Residual.
Lead (by difference) ..	98·79767	99·99297	23·97
Bismuth	0·00376	0·00305	11·20
Copper	0·37108	0·00060	14·44
Antimony	0·55641	0·00099	29·70
Silver	0·25400	,,	18·435
Tin	0·00575	0·00041	traces
Nickel..	0·00730	,,	0·090
Zinc	0·00271	0·00198	1·80
Sulphur	0·00132	,,	,,
	100·00000	100·00000	99·635

In his pamphlet, Mr. Keith says that his cost price will be
still more reduced, as Mr. Weston has undertaken to supply
him with a dynamo-electric machine for 1000 francs (40*l*.),
which will precipitate 40 tons of lead in twenty-four hours.
We have seen previously that 10 tons in twenty-four hours required 12 horse-power, the power required for this machine for
40 tons will therefore be 48 horse-power. But we know by
experience that such a dynamo costs to the manufacturer at
least 7500 francs (300*l*.), exclusive of general expenditure. Mr.

Weston enjoys, it is true, an excellent reputation in the United States; but if he has really accepted such an undertaking, he cannot, we believe, fail to be a heavy loser.

The consumption of electricity might perhaps be considerably reduced, which would allow of the use of much smaller dynamos; but this could only be done by using gigantic baths and treating some thousand tons of lead at a time, which would be a folly from an economical point of view. The interest of the supplementary capital required would be much greater than the profits derived from the economy in the motive power; the problem cannot therefore be solved by the adoption of small dynamo-electric machines.

The long explanations which we gave relatively to the treatment of copper equally apply to the treatment of lead, and to any metallic purification which it might be desired to electrolytically perform.

CHAPTER XIV.

TREATMENT OF ORES.

OF THE IMPORTANCE OF THE SUBJECT.—The use of electricity for the treatment of ores has already been made the subject of numerous patented combinations, and more or less practical experiments; but we believe that there does not exist one single industrial application based on the said use of electricity. We must, therefore, limit ourselves to the descriptions of those processes which appear to be the most ingenious and easy of realisation.

The importance of this problem is of the greatest moment, especially for the ores of precious metals, the actual metallurgy of which leaves somewhat to be desired. For example, the arsenio-sulphides and antimonides of silver and the auriferous pyrites, which are now so much neglected, and from which such a small yield is obtained, ought to attract the attention of electricians. The interest is much less for what concerns the gold and silver ores, which are easy of amalgamation; as to the copper, lead, and zinc ores, the cost of treating them by electricity will, whatever be done, be an almost insurmountable obstacle to its general use.

We have seen in the previous chapter how expensive the refining of copper proved to be; the operations are, however, extremely simple, and the work required in the electrolysis of the sulphate of copper with a soluble anode is very limited. If the cost of fuel and labour were not covered by the gold and silver obtained as sub-products, the pure coppers obtained from native copper would probably be used for telegraphic purposes, and those obtained by the ordinary metallurgic processes would be used for general purposes. The preparation of certain simple bodies is also one of the interesting branches of electro-metallurgy.

BUNSEN'S PROCESS.—Bunsen was the first who succeeded in rapidly preparing in comparatively large quantities the magnesium, the barium, the aluminium, and the calcium by means of electricity. He acted on some chlorides of these metals, either by dissolving them in water so as to obtain a concentrated solution, of which he raised the temperature at the same time that he submitted them to the decomposing action of the electric current, or in melting them in an anhydrous state in a porcelain crucible highly heated, and using as electrodes some well reduced coke carbon.*

For magnesium, for instance, Bunsen used well dried chloride of magnesium and placed it in a varnished porcelain crucible internally divided by means of a porous porcelain partition which did not reach the bottom; the porcelain cover was provided with two holes for the introduction of the conductors. As soon as the chloride reached a state of fusion, he sent through it a current of 15 to 20 volts, and the decomposition immediately took place: the chlorine going to the positive pole, and the magnesium to the negative. As the metal is not so dense as the chloride, the operator took care to retain it in the liquid by means of oblique grooves cut in the carbon electrode on which it was deposited; the magnesium would without this precaution have risen to the surface of the bath and been burned.

Bunsen prepared aluminium in the same manner, operating on a double chloride of ammonium and sodium (the chloride of aluminium does not melt, but vaporises at a low temperature).

M. Sainte-Claire Deville has perfected this latter process. To this effect he prepared the solution to be decomposed by mixing in a porcelain cup heated to about 200° C. 2 parts of

* De La Rive, vol. iii. pp. 511 and following.

chloride of aluminium and 4 parts of dried and pulverised sodic chloride. He filled with this solution and up to the same level, a varnished porcelain crucible and a well dried porous porcelain tube placed in the middle of the crucible. A platinum plate and a carbon plate were respectively used as negative and positive poles. The crucible was kept constantly heated. As soon as the electric current passed, the aluminium and the sodic chloride were deposited on the platinum plate, and the chlorine with a little chloride of aluminium evaporated. · The deposit was afterwards treated by water in order to dissolve the sodic chloride and the grey powder united into a regulus, by successive fusions, by means of the double chloride of aluminium and sodium as flux.

M. Sainte-Claire Deville afterwards prepared aluminium from sodium without using an electric current, and obtained a superior result with this second method which is outside our subject.

BECQUEREL'S PROCESS FOR THE GOLD, SILVER, AND COPPER ORES.—Becquerel more than thirty years ago described a complete method for the electric treatment of silver, lead, and copper ores; this method, ignored by many inventors, had the privilege of being from time to time patented under different names, and especially in countries where the legislature does not admit of an examination previous to the granting of the patent.

The complete description of the said process is to be found in the 'Traité d'électricité et de magnétisme,' by Messrs. C. and E. Becquerel, published in 1875. We will give a brief account of them.

The treatment of ores by the wet method, due to this celebrated scientist, is based on the property possessed by silver chloride and plumbic sulphate to dissolve in a saturated solution of sodic chloride and sulphate of copper in water.

The chloruretting of the silver is effected by the wet or dry method, according to the state of combination of the silver in the ore; the wet method is used when the silver is in a metallic state, or in a state of simple sulphide; the dry method is necessary if the silver is in a state of double sulphide.

To chloruret a silver ore, the first operation consists in

reducing it, by means of a crushing machine, to a state of impalpable powder, for the success of a good chloruretting almost solely depends on the state of extreme division of the ore. The substances used for chloruretting by the wet method are the sodium chloride, the sulphate of copper, or the sulphate of peroxide of iron, in proportions dependent on the composition of the ore and its percentage of silver. (The quantities required are approximately one-tenth in sodium chloride and one-twentieth in sulphate of the weight of the ore.)

The silver ore can also be chloruretted by previously mixing it with sodic chloride and gradually pouring 3 kilogrammes of nitric acid on 100 kilogrammes of the mixture.

The silver ores composed of double sulphides and of multiple combinations can only be chloruretted by the dry method, that is, by roasting, with addition of sodic chloride and often of pyrites. After the ore has been pulverised and mixed with the sea-salt and pyrites, the roasting is effected in a reverberatory furnace, stirring it constantly with a poker, so as to multiply the points of contact with air. The object of the roasting is to get rid of the sulphur, the arsenic, the selenium, and the antimony. A portion of the sulphur escapes in the state of sulphurous acid, and the other is transformed into sulphuric acid, which combines with the bases and forms sulphates. These sulphates react on the sodic chloride; there is formation of sodic sulphate and liberation of chlorine, which combines with silver.

Copper in ores is found in either a metallic state or in a state of oxide or chloride, of sulphide or carbonate, &c.; in the majority of cases its sulphide is combined with iron sulphide, and constitutes the cuprous pyrite; when combined with other sulphides it constitutes some multiple sulphides, the treatment of which by the dry method offers some difficulties. It is those combinations which are submitted to the roasting. When the copper is in a metallic state it is separated from the gangue by means of lixiviation; if it is in the state of oxide or carbonate, it must be combined with sulphuric acid; if it is in the state of simple, double, or multiple sulphide, it must be roasted with great care, so as not to produce any oxide. When the cuprous pyrite is roasted in a sustained but not too intense heat, the sulphide of iron which is reduced at a lower temperature changes into a state of peroxide, whereas the sulphate of copper is to be found in almost its totality in a state of anhydrous sulphate of copper in the roasted mass. When the ores contain oxide or carbonate of copper, iron pyrites must be added. In localities where sulphuric acid can be easily procured the cuprous ores composed of oxide or carbonate are directly transformed into sulphates.

As to lead, it should be sulphated by either the dry or the wet method. Plumbic sulphide requires less heat than galena, because it is fusible and volatile.

When galena is roasted, sulphurous acid is liberated, the lead is oxidised, and plumbic sulphate is formed. The sulphatation of galena is obtained by the wet method by making the sulphate of copper in solution react on the plumbic sulphide at the ordinary temperature by the intermediary of salt water.

All that relates to this first part of the operation is minutely described in M. Becquerel's work; it is the most important part, and that which is generally neglected now. It is forgotten that in an electrolytic operation the bath is the essential part, and it is particularly sought to economise the motive power. The result is nearly always a complete failure, whereas Becquerel has succeeded in rendering the electro-chemical operation a perfectly industrial one. His processes, truly, are expensive, but they actually exist, whereas the modern methods are very economical on paper, but cannot be realised in practice.

Silver, copper, and lead ores being chloruretted or sulphated, and the dissolutions of the metallic compounds being made in salt water for silver and lead, and in water for copper, these three metals can be obtained in a metallic state by the use of an electrical action arising from a simple bath or from an external source. For the electro-chemical treatment proper, Becquerel used voltaic couples composed of zinc, iron, or lead associated with copper bands, to tin plates, or to pieces of carbon placed in immediate contact with the metallic solution, whereas the oxidisable metals were placed in a porous cell filled with salt water. By grouping six simple baths so as to constitute a real battery of six cells, the chloride decomposed more rapidly without any increase of expenditure. In the first few hours three parts of the silver are extracted, the extraction of the remainder requiring a much longer time. This is due to the liquid, which becomes more resistant, and to the electric current, which decreases in strength.

For the treatment of copper ores Becquerel transformed the sulphides into sulphates with or without the assistance of sulphuric acid; the sulphate once obtained, was placed in a concentrated solution and submitted to the action of the electric current. More frequently he adopted the simple bath arrangement, using cast iron as a positive electrode; to this effect, he superposed two solutions in the same vessel, one saturated with denser sulphate of copper, the other with lighter sulphate of iron; in the first he introduced a copper strip, and in the second a cast-iron strip, communicating with the other strip by means of a metallic conductor. The couple resulting from that arrangement was sufficiently powerful to decompose the sulphate of copper; the oxygen and the acid of the sulphate were deposited on the cast iron forming sulphate of iron, whereas the copper was deposited on the copper strip.

This operation had a serious inconvenience; the deposited copper was first in a state of chemical purity, then it became brittle, friable, and charged with iron; in order to prevent these irregularities Becquerel conceived an apparatus which is worth describing in detail.

It consisted of a wooden case, lined with lead covered with wax, and intended for receiving the solution of sulphate of iron. This case was provided with two openings, the top one for the introduction of the normal liquor, the lower one for the expulsion of the too dense liquor by means of siphons. Some leaded sheet-iron boxes were immersed in the interior of this case, the end walls and bottom of which boxes were of metal, and the lateral walls of openwork lined with sheets of cardboard. An opening at the bottom admitted, by means of siphons, the concentrated solution of sulphate of copper, and

R

another opening at the top allowed the flowing in of the weak
solution. In one of these boxes was placed the metal intended
for the reception of the copper deposition, and between each of
them were the cast-iron plates intended for the production of
the current. The apparatus was regulated in such a manner,
that at every instant as much concentrated solution of copper
and weak solution of iron were admitted as weak copper liquor
and strong iron liquor run out; this action continued without
any manipulation. The only labour required consisted in taking
away the copper sheets when they were of a suitable thickness,
and replacing the cast-iron plates when they were used up.

Without stopping to consider the lead ore, which Becquerel
treated in a similar manner, we will quote, word for word, the
conclusion at which he arrived after ten years of ingenious
researches, experiments, and combinations respecting the electro-
chemical treatment of a great variety of ores.

BECQUEREL'S CONCLUSION.—" From the facts set forth in
this book," he says, in his 'Traité d'Électricité,' published in 1855,
" it evidently results that silver ores can be treated without
difficulty by the electro-chemical process when sea-salt is
at a low price and there is enough wood in the locality for
operating a roasting, if the chloruretting cannot be done by the
wet process; that this process is particularly applicable to very
complex ores, that is to say, sulphuretted ores; and that,
although it is simpler than the Mexican or freyberyan amalgamation, there are, nevertheless, instances in which it will be
preferred to this last method, which would not be suitable for
the treatment of argentiferous galena and argentiferous cuprous
pyrites. It is most probable that the electro-chemical method
by means of which silver, copper, and lead ores can be treated,
will be adopted in practice when the principles upon which it
rests shall have become familiar. It will more particularly
be adopted in countries where mercury is only found with
difficulty, and where wood is too scarce for treating the ore
by smelting and where common salt is abundant."

This conclusion, which nine years later was republished without any alteration by the author in his 'Traité d'Électro-chimie,'
can be thus summed up: the electro-chemical process for the

treatment of ores is good, but more expensive than the ordinary metallurgical process. It is exactly as if it were sought to obtain motive power by boring artesian wells; the engineer consulted would answer: This system may be adopted, particularly in countries where fuel is at an extraordinarily high price, and where no waterfalls are to be found.

It must be observed that Becquerel's works took place long before the invention of continuous-current dynamo-electric machines, notwithstanding which the question of production of electricity never jeopardised the success of his works. The factory which he established at Grenelle was well organised; the necessary electric currents were produced at a very moderate cost, and all the operations were well co-ordinated. If Becquerel has not obtained an industrial success, it is simply owing to the fact that the processes of amalgamation and roasting in ordinary use are more economical than the treatment by means of sodic chloride. Things were thus thirty-five years ago, and they have not altered since.

LAMBOTTE-DOUCET'S METHOD.—This process, applied to zinc ores, has been tried at the Bleyberg mines. It consists in dissolving the previously roasted ore in the hydrochloric acid of commerce so as to obtain a neutral and concentrated solution of zincic chloride. The iron is eliminated by means of chloride of lime and oxide of zinc; it precipitates in a state of ferric oxide. The chloride of zinc thus obtained is submitted to the action of the electric current with graphite anodes and zinc cathodes.

Under the action of the current the zinc is deposited on the cathode, and the chlorine is liberated at the anode. Under these conditions the anode polarises rapidly; the solution, originally neutral, becomes acid, and soon after the current does not precipitate any more metal.

LUCKOW'S METHOD.—This process is based on the use of the zinc ore as an anode. The electrolysis is done in large rectangular baths. The cathode is constituted of a zinc plate, and the anode of a mixture of ore and coke placed in an open-work case. A wooden frame loaded with lead and covered with a thick tissue is placed under the cathode in order to collect the precipitated zinc. The froth which, during the operation, is formed on the surface of the liquid must be skimmed from time to time.

When a solution of sulphate of zinc is used, great care must be taken to maintain the liquid in a neutral state; this is not so much necessary with chloride of zinc, but this latter solution has the inconvenience of generating chlorine at the anode. A suitable liquor, according to M. Luckow, for the direct extraction of the zinc from the blende, is a solution of sea-salt slightly acidulated.

LÉTRANGE'S METHOD.—This process, also applied to zinc ores, has been tried on a large scale at Romilly and at Saint-Denis, and consists, in principle,—

1st. In moderately roasting the blendes without overheating, so as to obtain zinc sulphate by directing the sulphurous vapours on to some oxidised ores (roasted blende, calcined calamines), in order also to transform them into sulphates.

2nd. In dissolving the sulphate of zinc thus obtained in order to make it into a concentrated solution.

3rd. In precipitating the metal of that solution, by using as anodes some plates of graphite, or of a metal which is not attacked by sulphuric acid.

The characteristic of the system is the use, as a dissolvent, of the sulphuric acid borrowed from the ore itself. The inventor deals with any ore, and particularly with those which are not much sought after, and are even abandoned with the usual processes. The preparation is an economical one, for it is not required to separate the plumbic substances, and in many cases the calcareous gangues may be left when the sulphuric acid produced in excess is to be absorbed in the treatment by reduction of the sulphate.

The calamine does not require to be calcined. The roasting of the blende must be conducted at a moderate temperature, so as to facilitate the formation of the greatest possible quantity of sulphate of zinc. The sulphurous vapours which are not retained in combination with the ore are brought into contact with the roasted blende and the calamine, for converting their zinc into sulphate or sulphite, which, exposed to the air, is soon sulphated.

A **small** proportion of blende used with calamine is sufficient for the supply of the acid consumed by the lime, the iron, and other foreign materials.

In the roasting of blende, **care must be** taken to avoid **its** being crushed, and the **crushing** should only be applied **to** the residual substances, which have remained refractory to **a** first attack and become more friable after being calcined.

After this summary preparation, the **ores are** transformed into sulphates and **put to** the electrolytic **treatment.** To that effect they **are placed** in large tanks, in which a **feeble current** of water dissolves **the** sulphate of zinc which has **been** formed. The sulphated **liquor is then** directed into a **series of tanks,** across **which it slowly travels,** depositing, **under the** action of the electric **current, a** portion **of the zinc** which it contains. The freed sulphuric acid **rises to the surface,** and runs out by overflowing. **The** acidulated **liquor is cast** over the heaps of **ores,** the acid dissolves the **oxides of** zinc, and the sulphate of **zinc is formed anew.** The **operation is** conducted in a continuous manner, **owing to** a difference of level between the tanks and **a** mechanical exertion **in one point** of the circuit.

The acid being **indefinitely** reproduced in the precipitating tanks, it **is sufficient that** the ores should contain a quantity **of** sulphate large enough **to** supply **the** acid which is absorbed **by** the foreign **matters** of the bath.

Lead, silver, **and other matters** insoluble **in** sulphuric acid **are** collected **in the** residuum and separately treated.

M. Létrange indicates **the following** process for transforming **the** blende into sulphate. **Without any** regard to the more or less complete transformation **of the** sulphurous acid vapours into sulphuric **acid,** and avoiding **the expenditure** of nitric acid, the **vapours produced** by **the** roasting are simply **made to pass** through **the** columns or chambers containing the **zinc ore,** which **is kept in a wet state by** a falling spray of water. The sulphurous acid forms soluble **zinc** sulphites, which become transformed into sulphates after **having** been exposed to the atmospheric air for a certain time.

M. Létrange **has** calculated that, with the processes actually in existence, the cost of installing **a** factory capable of annually treating 1,000,000 kilogrammes of zinc would be 1,000,000 francs, whereas it would not exceed 500,000 francs if acting by electrolysis and adopting his system. He has also calculated that with a judicious use of electricity he could attain a daily production of 10 to 12 kilogrammes of zinc per horse-power.

These calculations are somewhat premature and are of no industrial value, for they are based on doubtful hypotheses instead of precise and repeated trials.

The only verified experiment of M. Létrange's process was realised in 1882, by M. Cadiat, under the direction of our lamented friend Alfred Niaudet. It gave the following results:—

Number of baths in tension	5
Intensity of the current in amperes ..	75
Electromotive force in volts	13·05
Duration of the experiment	4h. 15m.
Weight of zinc obtained	1·475 kil.

The No. 2 Gramme machine used in this experiment had a mechanical efficiency of 0·75.

The expenditure of work was, therefore,—

$$\frac{75 \times 13 \cdot 05}{0 \cdot 75 \times 9 \cdot 81} = 133 \text{ kilogrammetres.}$$

Thus, to obtain 1·475 kilogramme of zinc, 133 kilogrammetres had to be consumed during 4 hours and 15 minutes, which corresponds to 565 kilogrammetres in 1 hour, and to 5 horse-power for 1 kilogramme per hour.

In order to ascertain the industrial efficiency of this experiment, it is necessary to determine the amount of motive power required for the dissociation of the zinc from its sulphate.

The heat of formation of 1 equivalent of sulphate of zinc is 53·5 calories.

One equivalent of sulphate of zinc contains 32·7 grammes of zinc. Therefore $\frac{53 \cdot 5 \times 1000}{32}$ calories will be required for the liberation of 1 kilogramme of zinc; this corresponds to 693,664 kilogrammetres, and to 2·6 horse-power during one hour.

Theoretically, 2·6 horse-power must liberate 1 kilogramme

of zinc; in the experiment previously described, 5 horse-power-hours were expended for obtaining the same result.

The efficiency was therefore $\frac{2\cdot6}{5} = 0\cdot52$. The yield is 4·8 kilogrammes per 24 hours per horse-power.

By perfecting his process, M. Létrange will no doubt obtain a better result, but he will certainly not exceed 6 to 7 kilogrammes per horse-power per 24 hours; still it will be necessary, in order to reach that result, to considerably reduce the resistance of his actual bath. From an electrical point of view everything turns on that: join the baths in series, give them immense surfaces of anodes and cathodes so as to reduce their resistance, and a great production with a minimum of dynamos and a small expenditure of motive power is obtained. Can this result be industrially obtained? We do not know; but what we are certain about is, that the success will never be impeded by the use of machines. Dynamos of considerable power, of simple working, and indefinite duration, may be economically established, but a superior efficiency to that indicated by the theory should not be expected from them; that is to say, the production of a great electrical energy without borrowing from any given source a still greater mechanical energy.

Messrs. Blas and Miest * when calculating the required power for decomposing sulphate of zinc by the current, found that theoretically 2½ horse-power were required for the production of 24 kilogrammes of zinc per day, which amount corresponds to 9½ kilogrammes per day for 1 horse-power. This calculation is correct, but Messrs. Blas and Miest are mistaken when they only value at 30 per cent. the practical efficiency of the electrical energy utilised in the precipitation of the metal, compared to the original expenditure of work. This efficiency can certainly reach as high a figure as 70 and even 80 per cent. M. Létrange therefore might, as we said above, hope to obtain 6 to 7 kilogrammes of zinc per horse-power, but never more; his own calculations, based, not on the

* 'Note sur l'application de l'électrolyse à la métallurgie,' by C. Blas and E. Miest.

theory, but on a comparison with Bunsen cells actuating electric motors, are inexact from the starting-point, and cannot therefore offer any guarantee as to certitude.

The only cause opposed to M. Létrange being able to reach to such a high degree of efficiency is the necessity which obliges him to use a current having an electromotive force exceeding 1·5 volt; this entails the decomposition of water and consequently an appreciable loss of the motive power.

The electromotive force required by the decomposition of sulphate of zinc effectively is $53·5 \times 0·0438 = 2·34$ volts. In his experiment, M. Cadiat had $13·05$ volts for 5 baths, or $2·6$ volts per bath. We do not know of any practical means for suppressing or even of lessening the effect of this abnormal decomposition.

TREATMENT OF THE SULPHIDES.—*Deligny's Method.*—We extract from a letter addressed by M. Deligny to the journal 'La Lumière Electrique,' in 1881, the following information respecting a mode of treating ores, which it appears to us has served as a basis to the processes of Messrs. Blas and Miest, and of M. Marchese, as we will explain further. M. Deligny writes:—"I started from this fact, that the various natural cupric sulphides and their compounds or mixtures with iron pyrites are tolerably good conductors of electricity, and are more or less rapidly attacked by an acid in the presence of nascent oxygen. It was therefore to be expected that one of these compounds taken as a positive electrode in any electrolytic action should throw off, at least slowly, a certain portion of its metal to the solution, from which the electrical action would afterwards easily withdraw it. In order to realise these conditions, I have placed in a rectangular vessel either a weak solution of sulphate of copper, or some ordinary acidulated water, such as is used in voltameters. I have then placed in the liquid a copper plate as a negative electrode, and for a positive electrode a carbon surrounded with pulverised copper ore, and contained either in a linen bag or in a porous cell, or even at the bottom of the electrolytic cell and without any diaphragm. With the electromotive force of one or two Bunsen cells, the reaction takes place rapidly enough, and

especially when the ores are not separated by a resisting diaphragm; in the case of the linen bag the action is immediate.

"I have successively operated on two qualities of ores.
The first one was a cupriferous iron pyrite containing 4·60
per cent. of copper, and in which the yellow copper pyrite
with the iron pyrite forms a very homogeneous mass. In the
second ore the cuprous pyrite formed little agglomerations or
spots disseminated in the mass; the percentage was 3·60 per
cent. of copper.

"After a few days' action of the battery, a notable proportion
of the two ores was attacked, and the copper resulting from it
was partially deposited on the negative electrode. But the
attack, as I had anticipated, was attended with different results
on the two ores. The first one, of homogeneous composition, had
had all its elements parallelly dissolved, so that the residue
still gave 4·57 per cent. of copper; the second ore, contrarily,
only contained 2·35 per cent. of copper after the attack. The
combination of cuprous pyrite with the iron pyrite had, therefore, been dissolved in a greater proportion than the martial
pyrite.

Blas and Miest.—In a note published in 1882, M. Blas,
Professor at the Louvain University, and M. Miest, Engineer,
made known a process for the treatment of sulphuretted ores,
which is not absolutely new, but has the merit of being well
presented.

"Our method," say the authors, "is based upon the following
facts :—

"1st. The natural metallic sulphides are in certain degrees
conductors of the galvanic current;

"2nd. The sulphuretted ores (mixtures of sulphides and
gangues) are conductors of the current, even when the proportion
of gangues is very large.

"3rd. If a solution of a salt, the acid of which attacks the
natural sulphides, is electrolysed, by using the latter as anodes
the metal of the sulphide is dissolved, whereas the sulphur
remains deposited on the anode. It is with the nitrates that
this operation is more easily conducted, and, in that case,
without the formation of sulphate."

The inventors quote, as an example, a bath of plumbic nitrate in which were a galena anode and an insoluble metal cathode.

The reactions which take place are the following ones:—

$$Pb\ (NO_6)^2 = Pb + (NO_6)^2.$$
$$PbS + (NO_6)^2 = Pb + (NO_6)^2 + S.$$

The lead under the action of the current goes to the cathode, and the acid on to the anode, where it attacks the galena and regenerates the plumbic nitrate. The bath remains constant and neutral, and may be indefinitely used. The acid of the salt escapes from the action of the current, since it only comes out of one combination to enter into another. The sulphur can easily be extracted, since it is separated. Polarisation is in a great proportion avoided owing to the absence of gaseous escapes.

Messrs. Blas and Miest's process comprises two distinct operations:—1st, the agglomeration of the ores; 2nd, the electrolysation.

The agglomeration is obtained by heat and compression. The ore, crushed into grains of about 5 millimetres diameter, is introduced into copper moulds and submitted to a pressure of about 100 atmospheres, then closed and heated in a furnace at about 600° C.; then pressed again on coming out of the furnace; and lastly, rapidly cooled for facilitating the taking to pieces of the plates.

Thus prepared, the ore plates are attached to some iron bars, which are connected by means of iron conductors to the positive pole of a dynamo-electric machine and suspended in the bath. The latter is formed by the solution of a neutral metallic salt suitable for the nature of the ore to be treated. For galena, the solution, as a matter of course, is plumbic nitrate; for blende it is the nitrate, sulphide or chloride of zinc, and so on. If the ore is compound, the solution is selected in accordance with the various metals which enter into its composition.

The cathodes are composed of plates of metal insoluble in

the solution and connected through iron conductors to the negative pole of the machine.

The baths are joined in tension **or** in quantity, more frequently in tension, so as **to** obtain the greatest possible metallic precipitation with a minimum of electricity.

Messrs. Blas and Miest, starting from this erroneous principle, that the maximum of utilisable electricity occurs when the external resistance is equal to the resistance of the machine itself, have made, on the motive power necessary for the treatment of a few **simple** ores, some calculations which are naturally incorrect. **We only** intend to rectify one out of some of the examples given by them in their pamphlet, and, in order to have a point of comparison with M. Létrange's process, **we** will select the treatment of blende.

Under the action of **the current**, sulphate of zinc is decomposed into metallic zinc, which **is** deposited on the cathode, and sulphuric acid, which attacks the blende of the anode, dissolves it with formation **of** sulphate of zinc and sulphur deposition, as indicated in the following equations:—

$ZnSO_4 + $ galvanic current $ = Zn + SO_4$.

Work absorbed : **53·5 calories.**

$ZnS + SO_4 = ZnSO_4 + S$.

Work produced : $53·5 - 21·5$ calories.

The heat **of** formation of **ZnS being 21·5** calories, the chemical work of electricity **is** therefore **21·5** calories per equivalent, **or** 658 calories per kilogramme **of** zinc, which corresponds, per kilogramme of precipitated zinc per hour, to

$$\frac{21500 \times 424}{32·7 \times 75 \times 60 \times 60} = 1·04.$$

Theoretically, 1 horse-power **is** sufficient for precipitating **1** kilogramme of zinc per hour; in practice, if a possible efficiency of 70 per cent. is admitted, 1·49 horse-power will be required; whereas, according to the inventors, the maximum efficiency being only 30 per cent., 3·47 horse-power would be required. With 1 horse-power, 19 kilogrammes of zinc could therefore be precipitated in 24 hours by means of the Blas and

Miest process. We have seen that by the Létrange process, with the same power and in the same time, only 7 kilogrammes could be precipitated. If things occurred exactly as the inventors would have it, there would therefore be a great advantage in using agglomerate ores; unhappily, it is scarcely possible to maintain the electrolyte at a certain degree of saturation and composition, and the results are far from being satisfactory. We believe that this treatment also has never left the field of experiment and expectation to enter into that of a really industrial practice.

The principal defect of the preceding method is in the neglect with which the composition of the electrolyte is effected. In order to economise the motive power, Messrs. Blas and Miest have taken soluble anodes without submitting them to any preliminary chemical preparation. These anodes, containing several soluble substances foreign to the body which they had to precipitate, rapidly disturb the composition of the bath. The laboratory experiments being particularly directed towards the observation of the attack of the anode and the metallic precipitation on the cathode, have been successful enough; but when the inventors attempted to produce on an industrial scale they met with insuperable difficulties.

The fact that it is the bath and not the anode which has to be electrolysed should never be lost sight of. All the exertions of the manufacturer must be concentrated upon that point: ensure the constancy of the bath, and if they cannot obtain it without incurring too considerable an expenditure, the best plan is to abandon the electrolytic treatment and go back to the ordinary metallurgical treatment.

But before adopting such an extreme decision, they must attempt to combine the two modes of treatment. For example, instead of using the copper ores in their natural state, they might be first reduced into mattes of an average percentage. The purer pyrite $Cu_2S + Fe_2S_3$ contains 35 per cent. of copper, but it is always mixed with iron pyrite which considerably reduces its percentage, the latter rarely reaching more than 7 to 8 per cent. An ore composed of sulphides, of sulphates and oxides of the two metals, with a siliceous gangue, may be obtained by means of a partial roasting. If these materials are smelted together, a matte containing all the sulphur, almost all the copper, and a portion of the iron is formed. This matte can in its turn be treated by a partial roasting, followed by a smelting with an addition of siliceous substances. By these means, and after a certain number of not at all onerous operations, a matte sufficiently rich in copper for being used as an anode in an electrolytic bath could be obtained.

We are perfectly aware that our advice amounts to proposing the use of the current for refining the copper instead of using it for the treatment of natural sulphides; this is true, but it is a logical consequence of the theory of electrolysis, and it is, we believe, impossible to otherwise operate in an industrial installation.

Marchese's Method.—If we admitted that M. Marchese applies in the works of the Società di Miniere di Rame e di Electro-metallurgia, near Genoa, the process for which he took out a patent in 1883, and which is essentially based on the anodes obtained by the compression of the copper ore in metallic cases, we should have to alter our opinion, for the results are, it appears, satisfactory. But we know from a letter, published by the Director of the works in ‘La Lumière Électrique,’ that, with the view of notably reducing the resistance of the circuit in the electrolytic baths, the mode of formation of the anodes has been altered.

It is useless to discuss the object of this change; it suffices us to know that the anodes are no longer formed with natural ores without roasting and a previous fusion, to be satisfied and to believe in the financial success of the undertaking, and particularly if the ores contain small quantities of gold and silver.

Lambert's Method.—This process particularly relates to the treatment of gold and silver ores. The ore is dissolved by nascent chlorine, the latter being itself obtained by the decomposition of a soluble chloride under the action of an electric current. The metals of the ore are thereby transformed into chloride, and are dissolved in the bath, either in virtue of their own solubility, or owing to the salts which enter into the composition of the baths. **An ulterior** electrical action decomposes these chlorides in their turn, and gives metallic deposits on the cathode.

In order to obtain some result, even **small, with this** process, the following conditions must be observed :—

1st. Polarisation at the **anode and at the cathode must be** avoided. Polarisation is counteracted **by giving a continuous** motion to the mass of ore, which liberates the adhering gas. Continuous collection of the deposits at the cathode prevents polarisation.

2nd. In order to obtain a **complete attack of the ore, it** must be submitted, in all its parts, to the action of the nascent chlorine. This result is obtained **by giving the** mass a continuous motion, thus bringing all its points into contact with the conductors of electricity.

The experimental apparatus used by Mr. Lambert, consists of a case divided into two compartments by means of a porous partition. One of these compartments receives the bath in which the cathode is immersed; the cathode being a simple copper strip, which is cleaned **and replaced at regular in-**tervals of time. The ore is placed in the other compartment. Electric conductivity is established by means of a carbon plate facing the cathode, and upon which rest some transversal partitions, also of carbon material ; so that the ore is maintained in a veritable carbon carcase. The stirring is obtained by means of a current of water specially adapted for the purpose.

Werdermann's Method.—Amongst the numerous patents taken out for the electro-chemical treatment of metallic ores, we shall mention those of Werdermann, which made a certain sensation in the United States.

Werdermann, in the first instance, sent a very powerful current into the furnaces when the ores were in fusion, or at least at a red heat. Then, after some absolutely fruitless experiments, he decided upon a more rational, if not more industrial treatment.

This treatment was as follows :—

The ores containing some precious metals associated with sulphur, arsenic, antimony, or other analogous substances, were first submitted to a preliminary oxidisation by means of ozone. After oxidisation and lixiviation, the silver was deposited from the solution by electrical precipitation. The gold and silver remaining in the deposit were submitted to amalgamation. The latter was facilitated by wetting the ore with a solution of caustic alkali, and by connecting the stirring apparatus of the amalgamating vessel to the negative pole of a battery having its positive pole connected to the vessel itself. Werdermann has also described a process which consisted in roasting the ore in a current of vapours of sodic chloride, of air, and oxygen at the same time. The chlorides formed were dissolved by electricity.

Cobley's Method.—The sulphuretted ores, and especially copper ores, are roasted so as to obtain sulphate of copper, which is dissolved in water; the iron is precipitated by means of lime, and the solution submitted to electrolysis. In order to prevent the polarisation due to the liberation of oxygen on the insoluble anode, the inventor proposes to send into the bath a current of sulphurous acid generated by the roasting. The acid combines with the oxygen and there is a production of sulphuric acid, which remains dissolved.

We cannot recommend processes so summarily described by their inventors; we only insert them here in order to direct young experimenters, and prevent them wasting their time upon methods already unsuccessfully attempted by their predecessors.

TRIAL OF THE TREATMENT OF ARGENTIFEROUS COPPER ORES.—We will record an interesting trial made in Japan in 1882, on argentiferous ores. The operation was conducted with the solution of double chloride of silver and sodium.

The solution was placed in two glass vessels connected by a tube, the extremities of which lined with a linen cloth did not allow of the passage of any solid particle.

Two electrodes of platinum wire were next introduced in the vessels. The chlorine, which is rendered free by the decomposition of the dissolved chloride, attacks the copper of the pyrite, giving off cupric chloride, iron perchloride, as well as sulphate of copper and of iron.

These two last named **salts are,** under the action of the sodic chloride, transformed into cupric chloride, **iron** perchloride, and sodic **sulphate.**

The **decomposition** is maintained by allowing a fresh solution of sodic chloride to fall drop by drop into the vessels.

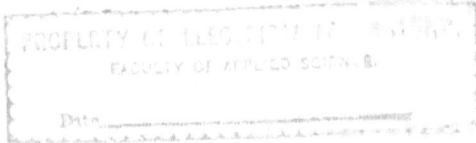

LONDON: PRINTED BY WILLIAM CLOWES AND SONS, LIMITED,
STAMFORD STREET AND CHARING CROSS.

1889.

BOOKS RELATING

TO

APPLIED SCIENCE,

PUBLISHED BY

E. & F. N. SPON,

LONDON: 125, STRAND.

NEW YORK: 12, CORTLANDT STREET.

The Engineers' Sketch-Book of Mechanical Movements, Devices, Appliances, Contrivances, Details employed in the Design and Construction of Machinery for every purpose. Collected from numerous Sources and from Actual Work. Classified and Arranged for Reference. *Nearly* 2000 *Illustrations.* By T W. BARBER, Engineer. 8vo, cloth, 7s. 6d.

A Pocket-Book for Chemists, Chemical **Manufacturers,** *Metallurgists, Dyers, Distillers,* **Brewers, Sugar Refiners,** *Photographers, Students, etc., etc.* By THOMAS BAYLEY, Assoc. R.C. Sc. Ireland, Analytical and Consulting Chemist and Assayer. **Fourth** edition, **with** additions, 437 pp., royal 32mo, roan, **gilt edges,** 5s.

SYNOPSIS OF CONTENTS :

Atomic Weights **and** Factors—Useful Data—Chemical Calculations—Rules for Indirect Analysis—Weights **and** Measures — Thermometers and Barometers — Chemical Physics — Boiling Points, etc.—Solubility of Substances—Methods of Obtaining Specific Gravity—Conversion of Hydrometers—Strength of Solutions by Specific Gravity—Analysis—Gas Analysis— Water Analysis—Qualitative Analysis and Reactions—Volumetric Analysis—Manipulation— Mineralogy — Assaying — Alcohol — Beer — Sugar — Miscellaneous Technological matter relating to Potash, Soda, Sulphuric Acid, Chlorine, **Tar** Products, Petroleum, Milk, Tallow, Photography, Prices, Wages, Appendix, etc., **etc.**

The Mechanician : **A Treatise on the** Construction and Manipulation of Tools, for the use and instruction of Young Engineers and Scientific Amateurs, comprising the Arts of Blacksmithing and Forging ; the Construction and Manufacture of Hand Tools, and the various Methods of Using and Grinding them ; the Construction of Machine Tools, and how to work them ; Machine Fitting and Erection ; description of Hand and Machine Processes ; Turning and Screw Cutting ; principles of Constructing and details of Making and Erecting Steam Engines, and the various details of setting out work, etc., etc. By CAMERON KNIGHT, Engineer. *Containing* 1147 *illustrations,* and 397 pages of letter-press, Fourth edition, 4to, cloth, 18s.

B

Just Published, in Demy 8vo, cloth, containing 975 pages and 250 Illustrations, price 7s. 6d.

SPONS' HOUSEHOLD MANUAL:
A Treasury of Domestic Receipts and Guide for Home Management.

PRINCIPAL CONTENTS.

Hints for selecting a good House, pointing out the essential requirements for a good house as to the Site, Soil, Trees, Aspect, Construction, and General Arrangement; with instructions for Reducing Echoes, Waterproofing Damp Walls, Curing Damp Cellars.

Sanitation.—What should constitute a good Sanitary Arrangement; Examples (with illustrations) of Well- and Ill-drained Houses; How to Test Drains; Ventilating Pipes, etc.

Water Supply.—Care of Cisterns; Sources of Supply; Pipes; Pumps; Purification and Filtration of Water.

Ventilation and Warming.—Methods of Ventilating without causing cold draughts, by various means; Principles of Warming; Health Questions; Combustion; Open Grates; Open Stoves; Fuel Economisers; Varieties of Grates; Close-Fire Stoves; Hot-air Furnaces; Gas Heating; Oil Stoves; Steam Heating; Chemical Heaters; Management of Flues; and Cure of Smoky Chimneys.

Lighting.—The best methods of Lighting, Candles, Oil Lamps, Gas, Incandescent Gas, Electric Light; How to test Gas Pipes; Management of Gas.

Furniture and Decoration.—Hints on the Selection of Furniture; on the most approved methods of Modern Decoration; on the best methods of arranging Bells and Calls; How to Construct an Electric Bell.

Thieves and Fire.—Precautions against Thieves and Fire; Methods of Detection; Domestic Fire Escapes; Fireproofing Clothes, etc.

The Larder.—Keeping Food fresh for a limited time; Storing Food without change, such as Fruits, Vegetables, Eggs, Honey, etc.

Curing Foods for lengthened Preservation, as Smoking, Salting, Canning, Potting, Pickling, Bottling Fruits, etc.; Jams, Jellies, Marmalade, etc.

The Dairy.—The Building and Fitting of Dairies in the most approved modern style; Butter-making; Cheesemaking and Curing.

The Cellar.—Building and Fitting; Cleaning Casks and Bottles; Corks and Corking; Aërated Drinks; Syrups for Drinks; Beers; Bitters; Cordials and Liqueurs; Wines; Miscellaneous Drinks.

The Pantry.—Bread-making; Ovens and Pyrometers; Yeast; German Yeast; Biscuits; Cakes; Fancy Breads; Buns.

The Kitchen.—On Fitting Kitchens; a description of the best Cooking Ranges, close and open; the Management and Care of Hot Plates, Baking Ovens, Dampers, Flues, and Chimneys; Cooking by Gas; **Cooking by Oil;** the Arts of Roasting, Grilling, Boiling, Stewing, Braising, Frying.

Receipts for Dishes—Soups, Fish, Meat, Game, Poultry, Vegetables, Salads, Puddings, Pastry, Confectionery, Ices, etc., etc.; Foreign Dishes.

The Housewife's Room.—Testing Air, Water, and Foods; Cleaning and Renovating; Destroying Vermin.

Housekeeping, Marketing.

The Dining-Room.—Dietetics; Laying and Waiting at Table; Carving; Dinners, Breakfasts, Luncheons, Teas, Suppers, etc.

The Drawing-Room.—Etiquette; Dancing; Amateur Theatricals; Tricks and Illusions; Games (indoor).

The Bedroom and Dressing-Room; Sleep; the Toilet; Dress; Buying Clothes; Outfits; Fancy Dress.

The Nursery.—The Room; Clothing; Washing; Exercise; Sleep; Feeding; Teething; Illness; Home Training.

The Sick-Room.—The Room; the Nurse; the Bed; Sick Room Accessories; Feeding Patients; Invalid Dishes and Drinks; Administering Physic; Domestic Remedies; Accidents and Emergencies; Bandaging; Burns; Carrying Injured Persons; Wounds; Drowning; Fits; Frost-bites; Poisons and Antidotes; Sunstroke; Common Complaints; Disinfection, etc.

The Bath-Room.—Bathing in General; Management of Hot-Water System.

The Laundry.—Small Domestic Washing Machines, and methods of getting up linen; Fitting up and Working a Steam Laundry.

The School-Room.—The Room and its Fittings; **Teaching, etc.**

The Playground.—Air and Exercise; Training; Outdoor **Games and Sports.**

The Workroom.—Darning, Patching, and Mending Garments.

The Library.—Care of Books.

The Garden.—Calendar of Operations **for Lawn, Flower** Garden, and Kitchen Garden.

The Farmyard.—Management of **the** Horse, Cow, Pig, Poultry, Bees, etc., etc.

Small Motors.—A description **of the** various small Engines useful for domestic purposes, from 1 man to 1 horse power, worked by various methods, such as Electric Engines, Gas Engines, Petroleum Engines, Steam Engines, Condensing Engines, Water Power, Wind Power, and the various methods of working and managing them.

Household Law.—The Law relating to Landlords and Tenants, Lodgers, Servants, Parochial Authorities, Juries, Insurance, Nuisance, etc.

On Designing Belt Gearing. By E. J. COWLING
WELCH, Mem. Inst. Mech. Engineers, Author of 'Designing Valve Gearing.' Fcap. 8vo, sewed, 6d.

A Handbook of Formulæ, Tables, and Memoranda,
for Architectural Surveyors and **others engaged in** Building. By J. T. HURST, C.E. Fourteenth edition, **royal 32mo,** roan, 5s.

"It is no disparagement to the many excellent publications we refer to, to say that in our opinion this little pocket-book of Hurst's is the very best of them all, without any exception. It would be useless to attempt a recapitulation of the contents, for it appears to contain almost *everything* that anyone connected with building could require, and, best of all, made up in a compact form for carrying in the pocket, measuring only 5 in. by 3 in., and about ⅜ in. thick, in a limp cover. We congratulate the author on the success of his laborious and practically compiled little book, which has received unqualified and deserved praise from every professional person to whom we have shown it."— *The Dublin Builder.*

Tabulated Weights of Angle, Tee, Bulb, Round,
Square, and Flat Iron **and** Steel, **and other** information for **the use of** Naval Architects and Shipbuilders. **By C. H.** JORDAN, M.I.N.A. **Fourth** edition, 32mo, **cloth, 2s.** 6d.

A Complete Set of Contract Documents for a Country
Lodge, comprising Drawings, Specifications, Dimensions (for quantities), Abstracts, Bill of Quantities, Form of Tender and Contract, with Notes by J. LEANING, printed in facsimile of the original documents, **on single** sheets fcap., in paper case, 10s.

A Practical Treatise on Heat, as applied to **the**
Useful Arts; for the Use of Engineers, Architects, &c. By THOMAS BOX. With 14 *plates*. Third edition, crown 8vo, cloth, 12s. 6d.

A Descriptive Treatise on Mathematical Drawing
Instruments: their construction, uses, qualities, selection, preservation, and suggestions for improvements, with hints upon Drawing and Colouring. By W. F. STANLEY, M.R.I. Fifth edition, *with numerous illustrations,* crown 8vo, cloth, 5s.

Quantity Surveying. By J. LEANING. With 42 illustrations. Second edition, revised, crown 8vo, cloth, 9s.

CONTENTS :

A complete Explanation of the London Practice.
General Instructions.
Order of Taking Off.
Modes of Measurement of the various Trades.
Use and Waste.
Ventilation and Warming.
Credits, with various Examples of Treatment.
Abbreviations.
Squaring the Dimensions.
Abstracting, with Examples in illustration of each Trade.
Billing.
Examples of Preambles to each Trade.
Form for a Bill of Quantities.
    Do.   Bill of Credits.
    Do.   Bill for Alternative Estimate.
Restorations and Repairs, and Form of Bill.
Variations before Acceptance of Tender.
Errors in a Builder's Estimate.

Schedule of Prices.
Form of Schedule of Prices.
Analysis of Schedule of Prices.
Adjustment of Accounts.
Form of a Bill of Variations.
Remarks on Specifications.
Prices and Valuation of Work, with Examples and Remarks upon each Trade.
The Law as it affects Quantity Surveyors, with Law Reports.
Taking Off after the Old Method.
Northern Practice.
The General Statement of the Methods recommended by the Manchester Society of Architects for taking Quantities.
Examples of Collections.
Examples of " Taking Off" in each Trade.
Remarks on the Past and Present Methods of Estimating.

Spons' Architects' and Builders' Price Book, with *useful Memoranda.* Edited by W. YOUNG, Architect. Crown 8vo, cloth, red edges, 3s. 6d. *Published annually.* Sixteenth edition. *Now ready.*

Long-Span Railway Bridges, comprising Investigations of the Comparative Theoretical and Practical Advantages of the various adopted or proposed Type Systems of Construction, with numerous Formulæ and Tables giving the weight of Iron or Steel required in Bridges from 300 feet to the limiting Spans ; to which are added similar Investigations and Tables relating to Short-span Railway Bridges. Second and revised edition. By B. BAKER, Assoc. Inst. C.E. *Plates,* crown 8vo, cloth, 5s.

Elementary Theory and Calculation of Iron Bridges and *Roofs.* By AUGUST RITTER, Ph.D., Professor at the Polytechnic School at Aix-la-Chapelle. Translated from the third German edition, by H. R. SANKEY, Capt. R.E. With 500 *illustrations,* 8vo, cloth, 15s.

The Elementary Principles of Carpentry. By THOMAS TREDGOLD. Revised from the original edition, and partly re-written, by JOHN THOMAS HURST. Contained in 517 pages of letter-press, and *illustrated with 48 plates and 150 wood engravings.* Sixth edition, reprinted from the third, crown 8vo, cloth, 12s. 6d.

Section I. On the Equality and Distribution of Forces — Section II. Resistance of Timber — Section III. Construction of Floors — Section IV. Construction of Roofs — Section V. Construction of Domes and Cupolas — Section VI. Construction of Partitions — Section VII. Scaffolds, Staging, and Gantries — Section VIII. Construction of Centres for Bridges — Section IX. Coffer-dams, Shoring, and Strutting — Section X. Wooden Bridges and Viaducts — Section XI. Joints, Straps, and other Fastenings — Section XII. Timber.

The Builder's Clerk : a Guide to the Management of a Builder's Business. By THOMAS BALES. Fcap. 8vo, cloth, 1s. 6d.

Practical Gold-Mining: a Comprehensive Treatise
on the Origin and Occurrence of Gold-bearing Gravels, Rocks and Ores,
and the methods by which the Gold is extracted. By C. G. WARNFORD
LOCK, co-Editor of 'Gold, its Occurrence and Extraction.' *With 8 plates
and 271 engravings in the text*, super-royal 8vo, cloth, 2*l.* 2*s.*

Hot Water Supply: A Practical Treatise upon the
Fitting of Circulating Apparatus in connection with Kitchen Range and
other Boilers, to supply Hot Water for Domestic and General Purposes.
With a Chapter upon Estimating. *Fully illustrated*, crown 8vo, cloth, 3*s.*

Hot Water Apparatus: An Elementary Guide for
the Fitting and Fixing of Boilers and Apparatus for the Circulation of
Hot Water for Heating and for Domestic Supply, and containing a
Chapter upon Boilers and Fittings for Steam Cooking. 32 *illustrations*,
fcap. 8vo, cloth, 1*s.* 6*d.*

*The Use and Misuse, and the Proper and Improper
Fixing of a Cooking Range. Illustrated*, fcap. 8vo, sewed, 6*d.*

Iron Roofs: Examples of Design, Description. *Illus-*
trated with 64 Working Drawings of Executed Roofs. By ARTHUR T.
WALMISLEY, Assoc. Mem. Inst. C.E. Second edition, revised, imp. 4to,
half-morocco, 3*l.* 3*s.*

A History of Electric Telegraphy, to the Year 1837.
Chiefly compiled from Original Sources, and hitherto Unpublished Documents, by J. J. FAHIE, Mem. Soc. of Tel. Engineers, and of the International Society of Electricians, Paris. Crown 8vo, cloth, 9*s.*

Spons' Information for Colonial Engineers. Edited
by J. T. HURST. Demy 8vo, sewed.

No. 1, Ceylon. By ABRAHAM DEANE, C.E. 2*s.* 6*d.*
CONTENTS:
Introductory Remarks—Natural Productions—Architecture and Engineering—Topography, Trade, and Natural History—Principal Stations—Weights and Measures, etc., etc.

No. 2. Southern Africa, including the Cape Colony, Natal, and the
Dutch Republics. By HENRY HALL, F.R.G.S., F.R.C.I. With
Map. 3*s.* 6*d.* CONTENTS:
General Description of South Africa—Physical Geography with reference to Engineering
Operations—Notes on Labour and Material in Cape Colony—Geological Notes on Rock
Formation in South Africa—Engineering Instruments for Use in South Africa—Principal
Public Works in Cape Colony: Railways, Mountain Roads and Passes, Harbour Works,
Bridges, Gas Works, Irrigation and Water Supply, Lighthouses, Drainage and Sanitary
Engineering, Public Buildings, Mines—Table of Woods in South Africa—Animals used for
Draught Purposes—Statistical Notes—Table of Distances—Rates of Carriage, etc.

No. 3. India. By F. C. DANVERS, Assoc. Inst. C.E. With Map. 4*s.* 6*d.*
CONTENTS:
Physical Geography of India—Building Materials—Roads—Railways—Bridges—Irriga-
tion — River Works — Harbours — Lighthouse Buildings — Native Labour — The Principal
Trees of India—Money—Weights and Measures—Glossary of Indian Terms, etc.

Our Factories, Workshops, and Warehouses: their Sanitary and Fire-Resisting Arrangements. By B. H. THWAITE, Assoc. Mem. Inst. C.E. *With* 183 *wood engravings,* crown 8vo, cloth, 9s.

A Practical Treatise on Coal Mining. By GEORGE G. ANDRÉ, F.G.S., Assoc. Inst. C.E., Member of the Society of Engineers. *With* 82 *lithographic plates.* 2 vols., royal 4to, cloth, 3l. 12s.

A Practical Treatise on Casting and Founding, including descriptions of the modern machinery employed in the art. By N. E. SPRETSON, Engineer. Third edition, with 82 *plates* drawn to scale, 412 pp., demy 8vo, cloth, 18s.

The Depreciation of Factories and their Valuation. By EWING MATHESON, M. Inst. C.E. 8vo, cloth, 6s.

A Handbook of Electrical Testing. By H. R. KEMPE, M.S.T.E. Fourth edition, revised and enlarged, crown 8vo, cloth, 16s.

Gas Works: their Arrangement, Construction, Plant, and Machinery. By F. COLYER, M. Inst. C.E. *With* 31 *folding plates,* 8vo, cloth, 24s.

The Clerk of Works: a Vade-Mecum for all engaged in the Superintendence of Building Operations. By G. G. HOSKINS, F.R.I.B.A. Third edition, fcap. 8vo, cloth, 1s. 6d.

American Foundry Practice: Treating of Loam, Dry Sand, and Green Sand Moulding, and containing a Practical Treatise upon the Management of Cupolas, and the Melting of Iron. By T. D. WEST, Practical Iron Moulder and Foundry Foreman. Second edition, *with numerous illustrations,* crown 8vo, cloth, 10s. 6d.

The Maintenance of Macadamised Roads. By T. CODRINGTON, M.I.C.E, F.G.S., General Superintendent of County Roads for South Wales. 8vo, cloth, 6s.

Hydraulic Steam and Hand Power Lifting and Pressing Machinery. By FREDERICK COLYER, M. Inst. C.E., M. Inst. M.E. *With* 73 *plates,* 8vo, cloth, 18s.

Pumps and Pumping Machinery. By F. COLYER, M.I.C.E., M.I.M.E. *With* 23 *folding plates,* 8vo, cloth, 12s. 6d.

Pumps and Pumping Machinery. By F. COLYER. Second Part. *With* 11 *large plates,* 8vo, cloth, 12s. 6d.

A Treatise on the Origin, Progress, Prevention, and Cure of Dry Rot in Timber, with Remarks on the Means of Preserving Wood from Destruction by Sea-Worms, Beetles, Ants, etc. By THOMAS ALLEN BRITTON, late Surveyor to the Metropolitan Board of Works, etc., etc. *With* 10 *plates,* crown 8vo, cloth, 7s. 6d.

The Municipal and Sanitary Engineer's Handbook.

By H. PERCY BOULNOIS, Mem. Inst. C.E., Borough Engineer, Portsmouth. With numerous illustrations, demy 8vo, cloth, 12s. 6d.

CONTENTS:

The Appointment and Duties of the Town Surveyor—Traffic—Macadamised Roadways—Steam Rolling—Road Metal and Breaking—Pitched Pavements—Asphalte—Wood Pavements—Footpaths—Kerbs and Gutters—Street Naming and Numbering—Street Lighting—Sewerage—Ventilation of Sewers—Disposal of Sewage—House Drainage—Disinfection—Gas and Water Companies, etc., Breaking up Streets—Improvement of Private Streets—Borrowing Powers—Artirans' and Labourers' Dwellings—Public Conveniences—Scavenging, including Street Cleansing—Watering and the Removing of Snow—Planting Street Trees—Deposit of Plans—Dangerous Buildings—Hoardings—Obstructions—Improving Street Lines—Cellar Openings—Public Pleasure Ground—Cemeteries—Mortuaries—Cattle and Ordinary Markets—Public Slaughter-houses, etc.—Giving numerous Forms of Notices, Specifications, and General Information upon these and other subjects of great importance to Municipal Engineers and others engaged in Sanitary Work.

Metrical Tables. By G. L. MOLESWORTH, M.I.C.E.

32mo, cloth, 1s. 6d.

CONTENTS.

General—Linear Measures—Square Measures—Cubic Measures—Measures of Capacity—Weights—Combinations—Thermometers.

Elements of Construction for Electro-Magnets. By

Count TH. DU MONCEL, Mem. de l'Institut de France. Translated from the French by C. J. WHARTON. Crown 8vo, cloth, 4s. 6d.

Practical Electrical Units Popularly Explained, with

numerous illustrations and Remarks. By JAMES SWINBURNE, late of J. W. Swan and Co., Paris, late of Brush-Swan Electric Light Company, U.S.A. 18mo, cloth, 1s. 6d.

A Treatise on the Use of Belting for the Transmission of Power. By J. H. COOPER. Second edition, illustrated, 8vo, cloth. 15s.

A Pocket-Book of Useful Formulæ and Memoranda

for Civil and Mechanical Engineers. By GUILFORD L. MOLESWORTH, Mem. Inst. C.E., Consulting Engineer to the Government of India for State Railways. With numerous illustrations, 744 pp. Twenty-second edition, revised and enlarged, 32mo, roan, 6s.

SYNOPSIS OF CONTENTS:

Surveying, Levelling, etc.—Strength and Weight of Materials—Earthwork, Brickwork, Masonry, Arches, etc.—Struts, Columns, Beams, and Trusses—Flooring, Roofing, and Roof Trusses—Girders, Bridges, etc.—Railways and Roads—Hydraulic Formulæ—Canals, Sewers, Waterworks, Docks—Irrigation and Breakwaters—Gas, Ventilation, and Warming—Heat, Light, Colour, and Sound—Gravity: Centres, Forces, and Powers—Millwork, Teeth of Wheels, Shafting, etc.—Workshop Recipes—Sundry Machinery—Animal Power—Steam and the Steam Engine—Water-power, Water-wheels, Turbines, etc.—Wind and Windmills—Steam Navigation, Ship Building, Tonnage, etc.—Gunnery, Projectiles, etc.—Weights, Measures, and Money—Trigonometry, Conic Sections, and Curves—Telegraphy—Mensuration—Tables of Areas and Circumference, and Arcs of Circles—Logarithms, Square and Cube Roots, Powers—Reciprocals, etc.—Useful Numbers—Differential and Integral Calculus—Algebraic Signs—Telegraphic Construction and Formulæ.

Hints on Architectural Draughtsmanship. **By G. W.**
TUXFORD HALLATT. Fcap. 8vo, cloth, 1s. 6d.

Spons' Tables and Memoranda for Engineers;
selected and arranged by J. T. HURST, C.E., Author of 'Architectural
Surveyors' Handbook,' 'Hurst's Tredgold's Carpentry,' etc.   Ninth
edition, 64mo, roan, gilt edges, 1s.; or in cloth case, 1s. 6d.
This work is printed in a pearl type, and is so small, measuring only 2½ in. by 1½ in. by
⅜ in. thick, that it may be easily carried in the waistcoat pocket.

"It is certainly an extremely rare thing for a reviewer to be called upon to notice a volume
measuring but 2½ in. by 1⅝ in., yet these dimensions faithfully represent the size of the handy
little book before us. The volume—which contains 118 printed pages, besides a few blank
pages for memoranda—is, in fact, a true pocket-book, adapted for being carried in the waistcoat pocket, and containing a far greater amount and variety of information than most people
would imagine could be compressed into so small a space. . . . . The little volume has been
compiled with considerable care and judgment, and we can cordially recommend it to our
readers as a useful little pocket companion."—*Engineering.*

A Practical Treatise on Natural and Artificial
Concrete, its Varieties and Constructive Adaptations. By HENRY REID,
Author of the 'Science and Art of the Manufacture of Portland Cement.'
New Edition, *with* 59 *woodcuts and* 5 *plates,* 8vo, cloth, 15s.

Notes on Concrete and Works in Concrete; especially
written to assist those engaged upon Public Works. By JOHN NEWMAN,
Assoc. Mem. Inst. C.E., crown 8vo, cloth, 4s. 6d.

Electricity as a Motive Power. By Count TH. DU
MONCEL, Membre de l'Institut de France, and FRANK GERALDY, Ingé-
nieur des Ponts et Chaussées. Translated and Edited, with Additions, by
C. J. WHARTON, Assoc. Soc. Tel. Eng. and Elec. *With* 113 *engravings
and diagrams,* crown 8vo, cloth, 7s. 6d.

Treatise on Valve-Gears, with special consideration
of the Link-Motions of Locomotive Engines. By Dr. GUSTAV ZEUNER,
Professor of Applied Mechanics at the Confederated Polytechnikum of
Zurich. Translated from the Fourth German Edition, by Professor J. F.
KLEIN, Lehigh University, Bethlehem, Pa. *Illustrated,* 8vo, cloth, 12s. 6d.

The French-Polisher's Manual. By a French-
Polisher; containing Timber Staining, Washing, Matching, Improving,
Painting, Imitations, Directions for Staining, Sizing, Embodying,
Smoothing, Spirit Varnishing, French-Polishing, Directions for Re-
polishing. Third edition, royal 32mo, sewed, 6d.

Hops, their Cultivation, Commerce, and Uses in
various Countries. By P. L. SIMMONDS. Crown 8vo, cloth, 4s. 6d.

The Principles of Graphic Statics. By GEORGE
SYDENHAM CLARKE, Capt. Royal Engineers. *With* 112 *illustrations.*
4to, cloth, 12s. 6d.

Dynamo-Electric Machinery : A Manual for Students
of Electro-technics. By SILVANUS P. THOMPSON, B.A., D.Sc., Professor
of Experimental Physics in University College, Bristol, etc., etc. Third
edition, *illustrated*, 8vo, cloth, 16s.

Practical Geometry, Perspective, **and** *Engineering*
Drawing; a Course of Descriptive Geometry adapted to the Requirements of the Engineering Draughtsman, including the determination of
cast shadows and Isometric Projection, each chapter being followed by
numerous examples ; to which are added rules for Shading, Shade-lining,
etc., together with practical instructions as to the Lining, Colouring,
Printing, and general treatment of Engineering Drawings, with a chapter
on drawing Instruments. By GEORGE S. CLARKE, Capt. R.E. Second
edition, *with 21 plates.* **2** vols., cloth, **10s.** 6d.

The Elements of Graphic **Statics.** By Professor
KARL VON OTT, translated from **the German** by G. S. CLARKE, Capt.
R.E., Instructor in Mechanical **Drawing**, Royal Indian Engineering
College. *With 93 illustrations*, **crown 8vo,** cloth, 5s.

A Practical Treatise **on the Manufacture and Distri-**
bution *of Coal* **Gas.** By WILLIAM RICHARDS. **Demy 4to,** with *numerous*
wood engravings and 29 plates, cloth, 28s.

SYNOPSIS OF CONTENTS:

Introduction — History of Gas Lighting — Chemistry of Gas Manufacture, by Lewis
Thompson, Esq., M.R.C.S.—Coal, with Analyses, by J. Paterson, Lewis Thompson, and
G. R. Hislop, Esqrs.—Retorts, Iron and Clay—Retort Setting—Hydraulic Main—Con-
densers — Exhausters — Washers and Scrubbers — Purifiers — Purification — History of Gas
Holder — Tanks, Brick and Stone, Composite, Concrete, Cast-iron, Compound Annular
Wrought-iron — Specifications — Gas Holders — Station Meter — Governor — Distribution—
Mains—Gas Mathematics, or Formulæ for the Distribution of Gas, by Lewis Thompson, Esq.—
Services—Consumers' Meters—Regulators—Burners—Fittings—Photometer—Carburization
of Gas—Air Gas and Water Gas—Composition of Coal Gas, by Lewis Thompson, Esq.—
Analyses of Gas—Influence of Atmospheric Pressure and Temperature on Gas—Residual
Products—Appendix—Description of Retort Settings, Buildings, etc., etc.

The New Formula for Mean Velocity of Discharge
of *Rivers and Canals.* By W. R. KUTTER. Translated from articles in
the ' Cultur-Ingénieur,' by LOWIS D'A. JACKSON, Assoc. Inst. C.E.
8vo, cloth, 12s. 6d.

The Practical Millwright and Engineer's **Ready**
Reckoner; or Tables for finding the diameter and power of **cog-wheels,**
diameter, weight, and power of shafts, diameter and strength **of bolts, etc.**
By THOMAS DIXON. Fourth edition, 12mo, cloth, 3s.

Tin : Describing the Chief Methods of Mining,
Dressing and Smelting it abroad ; with Notes upon Arsenic, Bismuth and
Wolfram. By ARTHUR G. CHARLETON, Mem. American Inst. of
Mining Engineers. *With plates*, 8vo, cloth, 12s. 6d.

B 3

Perspective, Explained and Illustrated. **By G. S.**
CLARKE, Capt. R.E. *With illustrations,* 8vo, cloth, 3s. 6d.

Practical Hydraulics; a Series of Rules and Tables
for the use of Engineers, etc., etc. By THOMAS BOX. Fifth edition,
numerous plates, post 8vo, cloth, 5s.

The Essential Elements of Practical Mechanics;
based on the Principle of Work, designed for Engineering Students. By
OLIVER BYRNE, formerly Professor of Mathematics, College for Civil
Engineers. Third edition, *with* 148 *wood engravings,* post 8vo, cloth,
7s. 6d.

CONTENTS:

Chap. 1. How Work is Measured by a Unit, both with and without reference to a Unit
of Time—Chap. 2. The Work of Living Agents, the Influence of Friction, and introduces
one of the most beautiful Laws of Motion—Chap. 3. The principles expounded in the first and
second chapters are applied to the Motion of Bodies—Chap. 4. The Transmission of Work by
simple Machines—Chap. 5. Useful Propositions and Rules.

Breweries and Maltings: their Arrangement, Construction, Machinery, and Plant. By G. SCAMELL, F.R.I.B.A. Second
edition, revised, enlarged, and partly rewritten. By F. COLYER, M.I.C.E.,
M.I.M.E. *With* 20 *plates,* 8vo, cloth, 18s.

A Practical Treatise on the Construction of Hori-
zontal and Vertical Waterwheels, specially designed for the use of opera-
tive mechanics. By WILLIAM CULLEN, Millwright and Engineer. *With*
11 *plates.* Second edition, revised and enlarged, small 4to, cloth, 12s. 6d.

A Practical Treatise on Mill-gearing, Wheels, Shafts,
Riggers, etc.; for the use of Engineers. By THOMAS BOX. Third
edition, *with* 11 *plates.* Crown 8vo, cloth, 7s. 6d.

Mining Machinery: a Descriptive Treatise on the
Machinery, Tools, and other Appliances used in Mining. By G. G.
ANDRÉ, F.G.S., Assoc. Inst. C.E., Mem. of the Society of Engineers.
Royal 4to, uniform with the Author's Treatise on Coal Mining, con-
taining 182 *plates,* accurately drawn to scale, with descriptive text, in
2 vols., cloth, 3l. 12s.

CONTENTS:

Machinery for Prospecting, Excavating, Hauling, and Hoisting—Ventilation—Pumping—
Treatment of Mineral Products, including Gold and Silver, Copper, Tin, and Lead, Iron
Coal, Sulphur, China Clay, Brick Earth, etc.

Tables for Setting out Curves for Railways, Canals,
Roads, etc., varying from a radius of five chains to three miles. By A.
KENNEDY and R. W. HACKWOOD. *Illustrated,* 32mo, cloth, 2s. 6d.

The Science and Art of the Manufacture of Portland Cement, with observations on some of its constructive applications. *With 66 illustrations.* By HENRY REID, C.E., Author of 'A Practical Treatise on Concrete,' etc., etc. 8vo, cloth, 18s.

The Draughtsman's Handbook of Plan and Map Drawing; including instructions for the preparation of Engineering, Architectural, and Mechanical Drawings. *With numerous illustrations in the text, and 33 plates* (15 *printed in colours*). By G. G. ANDRÉ, F.G.S., Assoc. Inst. C.E. 4to, cloth, 9s.

CONTENTS:

The Drawing Office and its Furnishings—Geometrical Problems—Lines, Dots, and their Combinations—Colours, Shading, Lettering, Bordering, and North Points—Scales—Plotting —Civil Engineers' and Surveyors' Plans—Map Drawing—Mechanical and Architectural Drawing—Copying and Reducing Trigonometrical Formulæ, etc., etc.

The Boiler-maker's and Iron Ship-builder's Companion, comprising a series of original and carefully calculated tables, of the utmost utility to persons interested in the iron trades. By JAMES FODEN, author of ' Mechanical Tables,' etc. Second edition revised, *with illustrations,* crown 8vo, cloth, 5s.

Rock Blasting: a Practical Treatise on the means employed in Blasting Rocks for Industrial Purposes. By G. G. ANDRÉ, F.G.S., Assoc. Inst. C.E. *With 56 illustrations and* 12 *plates,* 8vo, cloth, 10s. 6d.

Painting and Painters' Manual: a Book of Facts for Painters and those who Use or Deal in Paint Materials. By C. L. CONDIT and J. SCHELLER. *Illustrated,* 8vo, cloth, 10s. 6d.

A Treatise on Ropemaking as practised in public and private Rope-yards, with a Description of the Manufacture, Rules, Tables of Weights, etc., adapted to the Trade, Shipping, Mining, Railways, Builders, etc. By R. CHAPMAN, formerly foreman to Messrs. Huddart and Co., Limehouse, and late Master Ropemaker to H.M. Dockyard, Deptford. Second edition, 12mo, cloth, 3s.

Laxton's Builders' and Contractors' Tables; for the use of Engineers, Architects, Surveyors, Builders, Land Agents, and others. Bricklayer, containing 22 tables, with nearly 30,000 calculations. 4to, cloth, 5s.

Laxton's Builders' and Contractors' Tables. Excavator, Earth, Land, Water, and Gas, containing 53 tables, with nearly 24,000 calculations. 4to, cloth, 5s.

Egyptian Irrigation. By W. WILLCOCKS, M.I.C.E., Indian Public Works Department, Inspector of Irrigation, Egypt. With Introduction by Lieut.-Col. J. C. Ross, R.E., Inspector-General of Irrigation. *With numerous lithographs and wood engravings*, royal 8vo, cloth, 1*l*. 16*s*.

Screw Cutting **Tables** *for Engineers and Machinists*, giving the values of the different trains of Wheels required to produce Screws of any pitch, calculated by Lord Lindsay, M.P., F.R.S., F.R.A.S., etc. Cloth, oblong, 2*s*.

Screw Cutting Tables, for the use of Mechanical Engineers, showing the proper arrangement of Wheels for cutting the Threads of Screws of any required pitch, with a Table for making the Universal Gas-pipe Threads and Taps. By W. A. MARTIN, Engineer. Second edition, oblong, cloth, 1*s*., or sewed, 6*d*.

A Treatise on a Practical Method of Designing Slide-Valve Gears by Simple Geometrical Construction, based upon the principles enunciated in Euclid's Elements, and comprising the various forms of Plain Slide-Valve and Expansion Gearing ; together with Stephenson's, Gooch's, and Allan's Link-Motions, as applied either to reversing or to variable expansion combinations. By EDWARD J. COWLING WELCH, Memb. Inst. Mechanical Engineers. Crown 8vo, cloth, 6*s*.

Cleaning and Scouring : a Manual for Dyers, Laundresses, and for Domestic Use. By S. CHRISTOPHER. 18mo, sewed, 6*d*.

A Glossary of Terms used in Coal Mining. By WILLIAM STUKELEY GRESLEY, Assoc. Mem. Inst. C.E., F.G.S., Member of the North of England Institute of Mining Engineers. *Illustrated with numerous woodcuts and diagrams*, crown 8vo, cloth, 5*s*.

A Pocket-Book for Boiler Makers and Steam Users, comprising a variety of useful information for Employer and Workman, Government Inspectors, Board of Trade Surveyors, Engineers in charge of Works and Slips, Foremen of Manufactories, and the general Steam-using Public. By MAURICE JOHN SEXTON. Second edition, royal 32mo, roan, gilt edges, 5*s*.

Electrolysis : a Practical Treatise on Nickeling, Coppering, Gilding, Silvering, the Refining of Metals, and the treatment of Ores by means of Electricity. By HIPPOLYTE FONTAINE, translated from the French by J. A. BERLY, C.E., Assoc, S.T.E. *With engravings.* 8vo, cloth, 9*s*.

Barlow's Tables of Squares, Cubes, Square Roots,
Cube Roots, Reciprocals of all Integer Numbers up to 10,000. Post 8vo,
cloth, 6s.

A Practical Treatise on the Steam Engine, containing Plans and Arrangements of Details for Fixed Steam Engines,
with Essays on the Principles involved in Design and Construction. By
ARTHUR RIGG, Engineer, Member of the Society of Engineers and of
the Royal Institution of Great Britain. Demy 4to, *copiously illustrated
with woodcuts and 96 plates,* in one Volume, half-bound morocco, 2l. 2s.;
or cheaper edition, cloth, 25s.

This work is not, in any sense, an elementary treatise, or history of the steam engine, but
is intended to describe examples of Fixed Steam Engines without entering into the wide
domain of locomotive or marine practice. To this end illustrations will be given of the most
recent arrangements of Horizontal, Vertical, Beam, Pumping, Winding, Portable, Semi-
portable, Corliss, Allen, Compound, and other similar Engines, by the most eminent Firms in
Great Britain and America. The laws relating to the action and precautions to be observed
in the construction of the various details, such as Cylinders, Pistons, Piston-rods, Connecting-
rods, Cross-heads, Motion-blocks, Eccentrics, Simple, Expansion, Balanced, and Equilibrium
Slide-valves, and Valve-gearing will be minutely dealt with. In this connection will be found
articles upon the Velocity of Reciprocating Parts and the Mode of Applying the Indicator,
Heat and Expansion of Steam Governors, and the like. It is the writer's desire to draw
illustrations from every possible source, and give only those rules that present practice deems
correct.

A Practical **Treatise** *on* **the** *Science of Land* **and**
Engineering Surveying, Levelling, Estimating Quantities, etc., with a
general description of the several Instruments required for Surveying,
Levelling, Plotting, etc. By H. S. MERRETT. Fourth edition, revised
by G. W. USILL, Assoc. Mem. Inst. C.E. 41 *plates, with illustrations
and tables,* royal 8vo, cloth, 12s. 6d.

PRINCIPAL CONTENTS :

Part 1. Introduction and the Principles of Geometry. Part 2. Land Surveying ; com-
prising General Observations—The Chain—Offsets Surveying by the Chain only—Surveying
Hilly Ground—To Survey an Estate or Parish by the Chain only—Surveying with the
Theodolite—Mining and Town Surveying—Railroad Surveying—Mapping—Division and
Laying out of Land—Observations on Enclosures—Plane Trigonometry. Part 3. Levelling—
Simple and Compound Levelling—The Level Book—Parliamentary Plan and Section—
Levelling with a Theodolite—Gradients—Wooden Curves—To Lay out a Railway Curve—
Setting out Widths. Part 4. Calculating Quantities generally for Estimates—Cuttings and
Embankments—Tunnels—Brickwork—Ironwork—Timber Measuring. Part 5. Description
and Use of Instruments in Surveying and Plotting—The Improved Dumpy Level—Troughton's
Level—The Prismatic Compass — Proportional Compass — Box Sextant—Vernier— Panta-
graph—Merrett's Improved Quadrant—Improved Computation Scale—The Diagonal Scale—
Straight Edge and Sector. Part 6. Logarithms of Numbers — Logarithmic Sines and
Co-Sines, Tangents and Co-Tangents—Natural Sines and Co-Sines—Tables for Earthwork,
for Setting out Curves, and for various Calculations, etc., etc., etc.

Health and Comfort **in** *House Building, or Ventila-*
tion with Warm Air by Self-Acting Suction Power, with Review of the
mode of Calculating the Draught in Hot-Air Flues, and with some actual
Experiments. By J. DRYSDALE, M.D., and J. W. HAYWARD, M.D.
Second edition, with Supplement, *with plates,* demy 8vo, cloth, 7s. 6d.

The Assayer's Manual: an Abridged Treatise on the Docimastic Examination of Ores and Furnace and other Artificial Products. By BRUNO KERL. Translated by W. T. BRANNT. *With* 65 *illustrations,* 8vo, cloth, 12s. 6d.

Dynamo - Electric Machinery: a Text - Book for Students of Electro-Technology. By SILVANUS P. THOMPSON, B.A., D.Sc., M.S.T.E. Third Edition, revised and enlarged, 8vo, cloth, 16s.

The Practice of Hand Turning in Wood, Ivory, Shell, *etc.,* with Instructions for Turning such Work in Metal as may be required in the Practice of Turning in Wood, Ivory, etc. ; also an Appendix on Ornamental Turning. (A book for beginners.) By FRANCIS CAMPIN. Third edition, *with wood engravings,* crown 8vo, cloth, 6s.

CONTENTS :

On Lathes—Turning Tools—Turning Wood—Drilling—Screw Cutting—Miscellaneous Apparatus and Processes—Turning Particular Forms—Staining—Polishing—Spinning Metals —Materials—Ornamental Turning, etc.

Treatise on Watchwork, *Past and Present.* By the Rev. H. L. NELTHROPP, M.A., F.S.A. *With* 32 *illustrations,* crown 8vo, cloth, 6s. 6d.

CONTENTS :

Definitions of Words and Terms used in Watchwork—Tools—Time—Historical Summary—On Calculations of the Numbers for Wheels and Pinions; their Proportional Sizes, Trains, etc.—Of Dial Wheels, or Motion Work—Length of Time of Going without Winding up—The Verge—The Horizontal—The Duplex—The Lever—The Chronometer—Repeating Watches—Keyless Watches—The Pendulum, or Spiral Spring—Compensation—Jewelling of Pivot Holes—Clerkenwell—Fallacies of the Trade—Incapacity of Workmen—How to Choose and Use a Watch, etc.

Algebra Self-Taught. By W. P. HIGGS, M.A., D.Sc., LL.D., Assoc. Inst. C.E., Author of ' A Handbook of the Differential Calculus,' etc. Second edition, crown 8vo, cloth, 2s. 6d.

CONTENTS :

Symbols and the Signs of Operation—The Equation and the Unknown Quantity— Positive and Negative Quantities—Multiplication—Involution—Exponents—Negative Exponents—Roots, and the Use of Exponents as Logarithms—Logarithms—Tables of Logarithms and Proportionate Parts — Transformation of System of Logarithms — Common Uses of Common Logarithms—Compound Multiplication and the Binomial Theorem—Division, Fractions, and Ratio—Continued Proportion—The Series and the Summation of the Series— Limit of Series—Square and Cube Roots—Equations—List of Formulæ, etc.

Spons' Dictionary of Engineering, *Civil,* *Mechanical,* *Military, and Naval;* with technical terms in French, German, Italian, and Spanish, 3100 pp., and *nearly* 8000 *engravings,* in super-royal 8vo, in 8 divisions, 5l. 8s. Complete in 3 vols., cloth, 5l. 5s. Bound in a superior manner, half-morocco, top edge gilt, 3 vols., 6l. 12s.

Notes in Mechanical Engineering. Compiled principally for the use of the Students attending the Classes on this subject at the City of London College. By HENRY ADAMS, Mem. Inst. M.E., Mem. Inst. C.E., Mem. Soc. of **Engineers.** Crown 8vo, cloth, 2s. 6d.

Canoe and Boat Building: a complete Manual **for** Amateurs, containing plain and comprehensive directions for the construction of Canoes, Rowing and Sailing Boats, and Hunting Craft. By W. P. STEPHENS. *With numerous illustrations and 24 plates of Working Drawings.* Crown 8vo, cloth, 7s. 6d.

Proceedings of the National Conference of Electricians, *Philadelphia,* October 8th to 13th, 1884. 18mo, cloth, **3s.**

Dynamo - Electricity, its **Generation, A**pplication, Transmission, Storage, and Measurement. **By G. B.** PRESCOTT. *With* 545 *illustrations.* 8vo, cloth, 1l. 1s.

Domestic *Electricity for* **Amateurs.** Translated from the French of E. HOSPITALIER, **Editor** of "L'Electricien," by C. J WHARTON, Assoc. Soc. Tel. Eng. *Numerous illustrations.* Demy 8vo, cloth, 9s.

CONTENTS:

1. Production of the Electric Current—2. Electric Bells—3. Automatic Alarms—4. Domestic Telephones—5. Electric Clocks—6. Electric Lighters—7. Domestic Electric Lighting—8. Domestic Application of the Electric Light—9. Electric Motors—10. Electrical Locomotion—11. Electrotyping, Plating, and Gilding—12. Electric Recreations—13. Various applications—Workshop of the Electrician.

Wrinkles **in** *Electric Lighting.* By VINCENT STEPHEN. *With* **illustrations.** 18mo, cloth, 2s. 6d.

CONTENTS:

1. The Electric Current and its production by Chemical means—2. Production of Electric Currents by Mechanical means—3. Dynamo-Electric Machines—4. Electric Lamps—5. Lead—6. Ship Lighting.

The Practical Flax Spinner; being a Description of the Growth, Manipulation, and Spinning of Flax and Tow. By LESLIE C. MARSHALL, of Belfast. *With illustrations.* 8vo, cloth, 15s.

Foundations and Foundation Walls for all classes of *Buildings,* Pile Driving, Building Stones and Bricks, Pier and Wall construction, Mortars, Limes, Cements, Concretes, Stuccos, &c. 64 *illustrations.* By G. T POWELL and F. BAUMAN. 8vo, cloth, 10s. 6d.

Manual for Gas Engineering Students.   By D. LEE.
18mo, cloth 1s.

Hydraulic Machinery, Past and Present. A Lecture
delivered to the London and Suburban Railway Officials' Association.
By H. ADAMS, Mem. Inst. C.E. *Folding plate.* 8vo, sewed, 1s.

Twenty Years with the Indicator. By THOMAS PRAY,
Jun., C.E., M.E., Member of the American Society of Civil Engineers.
2 vols., royal 8vo, cloth, 12s. 6d.

Annual Statistical Report of the Secretary to the
*Members of the Iron and Steel Association on the Home and Foreign Iron
and Steel Industries in* 1887. Issued March 1888. 8vo, sewed, 5s.

Bad Drains, and How to Test them ; with Notes on
the Ventilation of Sewers, Drains, and Sanitary Fittings, and the Origin
and Transmission of Zymotic Disease. By R. HARRIS REEVES. Crown
8vo, cloth, 3s. 6d.

Well Sinking. The modern practice of Sinking
and Boring Wells, with geological considerations and examples of Wells.
By ERNEST SPON, Assoc. Mem. Inst. C.E., Mem. Soc. Eng., and of the
Franklin Inst., etc. Second edition, revised and enlarged. Crown 8vo,
cloth, 10s. 6d.

The Voltaic Accumulator : an Elementary Treatise.
By ÉMILE REYNIER. Translated by J. A. BERLY, Assoc. Inst. E.E.
With 62 *illustrations,* 8vo, cloth, 9s.

List of Tests (Reagents), arranged in alphabetical
order, according to the names of the originators. Designed especially
for the convenient reference of Chemists, Pharmacists, and Scientists.
By HANS M. WILDER. Crown 8vo, cloth, 4s. 6d.

Ten Years' Experience in Works of Intermittent
Downward Filtration. By J. BAILEY DENTON, Mem. Inst. C.E.
Second edition, with additions. Royal 8vo, sewed, 4s.

A Treatise on the Manufacture of Soap and Candles,
Lubricants and Glycerin. By W. LANT CARPENTER, B.A., B.Sc. (late
of Messrs. C. Thomas and Brothers, Bristol). *With illustrations.* Crown
8vo, cloth, 10s. 6d.

The Stability of Ships explained simply, and calculated by a new Graphic method. By J. C. SPENCE, M.I.N.A. 4to, sewed, 3s. 6d.

Steam Making, or Boiler Practice. By CHARLES A. SMITH, C.E. 8vo, cloth, 10s. 6d.

CONTENTS:

1. The Nature of Heat and the Properties of Steam—2. Combustion.—3. Externally Fired Stationary Boilers—4. Internally Fired Stationary Boilers—5. Internally Fired Portable Locomotive and Marine Boilers—6. Design, Construction, and Strength of Boilers—7. Proportions of Heating Surface, Economic Evaporation, Explosions—8. Miscellaneous Boilers, Choice of Boiler Fittings and Appurtenances.

The Fireman's Guide; a Handbook on the Care of **Boilers.** By TEKNOLOG, föreningen T. I. Stockholm. Translated from **the third** edition, and revised by KARL P. DAHLSTROM, M.E. Second **edition.** Fcap. 8vo, cloth, 2s.

A Treatise on Modern Steam Engines and Boilers, including Land Locomotive, and Marine Engines and Boilers, for the use of Students. By FREDERICK COLYER, M. Inst. C.E., Mem. Inst. **M.E.** *With 36 plates.* 4to, cloth, 25s.

CONTENTS:

1. Introduction—2. Original Engines—3. Boilers—4. High-Pressure Beam Engines—5. Cornish Beam Engines—6. Horizontal Engines—7. Oscillating Engines—8. Vertical High-Pressure Engines—9. Special Engines—10. Portable Engines—11. Locomotive Engines—12. Marine Engines.

*Steam **Engine** Management;* a Treatise on the Working **and Management** of Steam Boilers. By F. COLYER, M. Inst. C.E., Mem. Inst. **M.E.** 18mo, cloth, 2s.

Land Surveying on the Meridian and Perpendicular System. By WILLIAM PENMAN, C.E. 8vo, cloth, 8s 6d.

The Topographer, his Instruments and Methods, designed for the use of Students, Amateur Photographers, Surveyors, Engineers, and all persons interested in the location and construction of works based upon Topography. *Illustrated with numerous plates, maps, and engravings.* By LEWIS M. HAUPT, A.M. 8vo, cloth, 18s.

A Text-Book of Tanning, embracing the Preparation of all kinds of Leather. By HARRY R. PROCTOR, F.C.S., of Low Lights Tanneries. *With illustrations.* Crown 8vo, cloth, 10s. 6d.

In super-royal 8vo, 1168 pp., *with* 2400 *illustrations*, in 3 Divisions, cloth, price 13*s.* 6*d.* each ; or 1 vol., cloth, 2*l.* ; or half-morocco, 2*l.* 8*s.*

A SUPPLEMENT

TO

SPONS' DICTIONARY OF ENGINEERING.

EDITED BY ERNEST SPON, MEMB. SOC. ENGINEERS.

Abacus, Counters, Speed Indicators, and Slide Rule.
Agricultural Implements and Machinery.
Air Compressors.
Animal Charcoal Machinery.
Antimony.
Axles and Axle-boxes.
Barn Machinery.
Belts and Belting.
Blasting. Boilers.
Brakes.
Brick Machinery.
Bridges.
Cages for Mines.
Calculus, Differential and Integral.
Canals.
Carpentry.
Cast Iron.
Cement, Concrete, Limes, and Mortar.
Chimney Shafts.
Coal Cleansing and Washing.

Coal Mining.
Coal Cutting Machines.
Coke Ovens. Copper.
Docks. Drainage.
Dredging Machinery.
Dynamo - Electric and Magneto-Electric Machines.
Dynamometers.
Electrical Engineering, Telegraphy, Electric Lighting and its practical details, Telephones
Engines, Varieties of.
Explosives. Fans.
Founding, Moulding and the practical work of the Foundry.
Gas, Manufacture of.
Hammers, Steam and other Power.
Heat. Horse Power.
Hydraulics.
Hydro-geology.
Indicators. Iron.
Lifts, Hoists, and Elevators.

Lighthouses, Buoys, and Beacons.
Machine Tools.
Materials of Construction.
Meters.
Ores, Machinery and Processes employed to Dress.
Piers.
Pile Driving.
Pneumatic Transmission.
Pumps.
Pyrometers.
Road Locomotives.
Rock Drills.
Rolling Stock.
Sanitary Engineering.
Shafting.
Steel.
Steam Navvy.
Stone Machinery.
Tramways.
Well Sinking.

London: E. & F. N. SPON, 125, Strand.
New York: 12, Cortlandt Street.

SPONS' ENCYCLOPÆDIA

OF THE

INDUSTRIAL ARTS, MANUFACTURES, AND COMMERCIAL PRODUCTS.

EDITED BY C. G. WARNFORD LOCK, F.L.S.

Among the more important of the subjects treated of, are the
following :—

Acids, 207 pp. 220 figs.
Alcohol, 23 pp. 16 figs.
Alcoholic Liquors, 13 pp.
Alkalies, 89 pp. 78 figs.
Alloys.   Alum.
Asphalt.   Assaying.
Beverages, 89 pp. 29 figs.
Blacks.
Bleaching Powder, 15 pp.
Bleaching, 51 pp. 48 figs.
Candles, 18 pp. 9 figs.
Carbon Bisulphide.
Celluloid, 9 pp.
Cements.   Clay.
Coal-tar Products, 44 pp.
14 figs.
Cocoa, 8 pp.
Coffee, 32 pp. 13 figs.
Cork, 8 pp. 17 figs.
Cotton Manufactures, 62
pp. 57 figs.
Drugs, 38 pp.
Dyeing and Calico
Printing, 28 pp. 9 figs.
Dyestuffs, 16 pp.
Electro-Metallurgy, 13
pp.
Explosives, 22 pp. 33 figs.
Feathers.
Fibrous Substances, 92
pp. 79 figs.
Floor-cloth, 16 pp. 21
figs.
Food Preservation, 8 pp.
Fruit, 8 pp.

Fur, **5 pp.**
Gas, **Coal, 8 pp.**
Gems.
Glass, 45 pp. 77 figs.
Graphite, 7 pp.
Hair, 7 pp.
Hair Manufactures.
Hats, 26 pp. 26 figs.
Honey.   Hops.
Horn.
Ice, 10 pp. 14 figs.
Indiarubber Manufac-
tures, 23 pp. 17 figs.
Ink, 17 pp.
Ivory.
Jute Manufactures, **11**
pp., 11 figs.
Knitted Fabrics —
Hosiery, 15 pp. 13 figs.
Lace, 13 pp. 9 figs.
Leather, 28 pp. 31 figs.
Linen Manufactures, 16
pp. 6 figs.
Manures, 21 pp. 30 figs.
Matches, 17 pp. 38 figs.
Mordants, 13 pp.
Narcotics, 47 pp.
Nuts, 10 pp.
Oils and Fatty Sub-
stances, 125 pp.
Paint.
Paper, 26 pp. 23 figs.
Paraffin, 8 pp. 6 figs.
Pearl and Coral, 8 pp.
Perfumes, 10 pp.

Photography, 13 pp. 20
figs.
Pigments, 9 pp. 6 figs.
Pottery, 46 pp. 57 figs.
Printing and Engraving,
20 pp. 8 figs.
Rags.
Resinous and Gummy
Substances, 75 pp. 16
figs.
Rope, **16** pp. 17 figs.
Salt, 31 pp. 23 figs.
Silk, 8 pp.
Silk Manufactures, 9 **pp.**
11 figs.
Skins, 5 pp.
Small Wares, 4 pp.
Soap and Glycerine, 39
pp. 45 figs.
Spices, 16 pp.
Sponge, 5 pp.
Starch, 9 pp. 10 figs.
Sugar, 155 pp. 134
figs.
Sulphur.
Tannin, **18 pp.**
Tea, **12 pp.**
Timber, 13 pp.
Varnish, 15 pp.
Vinegar, 5 pp.
Wax, 5 pp.
Wool, 2 pp.
Woollen Manufactures,
58 pp. 39 figs.

Crown 8vo, cloth, with illustrations, 5s.

WORKSHOP RECEIPTS,

FIRST SERIES.

By ERNEST SPON.

SYNOPSIS OF CONTENTS.

Bookbinding.
Bronzes and Bronzing.
Candles.
Cement.
Cleaning.
Colourwashing.
Concretes.
Dipping Acids.
Drawing Office Details.
Drying Oils.
Dynamite.
Electro - Metallurgy — (Cleaning, Dipping, Scratch-brushing, Batteries, Baths, and Deposits of every description).
Enamels.
Engraving on Wood, Copper, Gold, Silver, Steel, and Stone.
Etching and Aqua Tint.
Firework Making — (Rockets, Stars, Rains, Gerbes, Jets, Tourbillons, Candles, Fires, Lances, Lights, Wheels, Fire-balloons, and minor Fireworks).
Fluxes.
Foundry Mixtures.

Freezing.
Fulminates.
Furniture Creams, Oils, Polishes, Lacquers, and Pastes.
Gilding.
Glass Cutting, Cleaning, Frosting, Drilling, Darkening, Bending, Staining, and Painting.
Glass Making.
Glues.
Gold.
Graining.
Gums.
Gun Cotton.
Gunpowder.
Horn Working.
Indiarubber.
Japans, Japanning, and kindred processes.
Lacquers.
Lathing.
Lubricants.
Marble Working.
Matches.
Mortars.
Nitro-Glycerine.
Oils.

Paper.
Paper Hanging.
Painting in Oils, in Water Colours, as well as Fresco, House, Transparency, Sign, and Carriage Painting.
Photography.
Plastering.
Polishes.
Pottery—(Clays, Bodies, Glazes, Colours, Oils, Stains, Fluxes, Enamels, and Lustres).
Scouring.
Silvering.
Soap.
Solders.
Tanning.
Taxidermy.
Tempering Metals.
Treating Horn, Mother-o'-Pearl, and like substances.
Varnishes, Manufacture and Use of.
Veneering.
Washing.
Waterproofing.
Welding.

Besides Receipts relating to the lesser Technological matters and processes, such as the manufacture and use of Stencil Plates, Blacking, Crayons, Paste, Putty, Wax, Size, Alloys, Catgut, Tunbridge Ware, Picture Frame and Architectural Mouldings, Compos, Cameos, and others too numerous to mention.

London: **E. & F. N. SPON**, 125, Strand.
New York: 12, Cortlandt Street.

Crown 8vo, cloth, 485 pages, with illustrations, 5s.

WORKSHOP RECEIPTS,

SECOND SERIES.

By ROBERT HALDANE.

SYNOPSIS OF CONTENTS.

Pigments, Paint, and Painting: embracing the preparation of *Pigments*, including alumina lakes, blacks (animal, bone, Frankfort, ivory, lamp, sight, soot), blues (antimony, Antwerp, cobalt, cœruleum, Egyptian, manganate, Paris, Péligot, Prussian, smalt, ultramarine), browns (bistre, hinau, sepia, sienna, umber, Vandyke), greens (baryta, Brighton, Brunswick, chrome, cobalt, Douglas, emerald, manganese, mitis, mountain, Prussian, sap, Scheele's, Schweinfurth, titanium, verdigris, zinc), reds (Brazilwood lake, **carminated** lake, carmine, Cassius purple, cobalt pink, cochineal lake, colcothar, Indian red, madder lake, red chalk, red lead, vermilion), whites (alum, baryta, Chinese, lead sulphate, white lead—by American, Dutch, French, German, Kremnitz, and Pattinson processes, precautions in making, and composition of commercial samples—whiting, Wilkinson's white, zinc white), yellows (chrome, gamboge, Naples, orpiment, realgar, yellow lakes) ; *Paint* (vehicles, testing oils, driers, grinding, storing, applying, priming, drying, filling, coats, brushes, surface, water-colours, removing smell, discoloration ; miscellaneous paints—cement paint for carton-pierre, copper paint, gold paint, iron paint, lime paints, silicated paints, steatite paint, transparent paints, tungsten paints, window paint, zinc paints) ; *Painting* (general instructions, proportions of ingredients, measuring paint work ; carriage painting—priming paint, best putty, finishing colour, cause of cracking, mixing the paints, oils, driers, and colours, varnishing, importance of washing vehicles, re-varnishing, how to dry paint ; woodwork painting).

London: E. & F. N. SPON, 125, Strand.
New York: 12, Cortlandt Street.

Crown 8vo, cloth, 480 pages, with 183 illustrations, 5s.

WORKSHOP RECEIPTS,

THIRD SERIES.

By C. G. WARNFORD LOCK.

Uniform with the First and Second Series.

SYNOPSIS OF CONTENTS.

London: **E. & F. N. SPON, 125, Strand.**
New York: 12, Cortlandt Street.

WORKSHOP RECEIPTS,

FOURTH SERIES,

DEVOTED MAINLY TO HANDICRAFTS & MECHANICAL SUBJECTS.

By C. G. WARNFORD LOCK.

250 Illustrations, with Complete Index, and a General Index to the
Four Series, 5s.

Waterproofing — rubber goods, cuprammonium processes, miscellaneous
preparations.

Packing and Storing articles of delicate odour or colour, of a deliquescent
character, liable to ignition, apt to suffer from insects or damp, or easily
broken.

Embalming and Preserving anatomical specimens.

Leather Polishes.

Cooling Air and Water, producing low temperatures, making ice, cooling
syrups and solutions, and separating salts from liquors by refrigeration.

Pumps and Siphons, embracing every useful contrivance for raising and
supplying water on a moderate scale, and moving corrosive, tenacious,
and other liquids.

Desiccating—air- and water-ovens, and other appliances for drying natural
and artificial products.

Distilling—water, tinctures, extracts, pharmaceutical preparations, essences,
perfumes, and alcoholic liquids.

Emulsifying as required by pharmacists and photographers.

Evaporating—saline and other solutions, and liquids demanding special
precautions.

Filtering—water, and solutions of various kinds.

Percolating and Macerating.

Electrotyping.

Stereotyping by both plaster and paper processes.

Bookbinding in all its details.

Straw Plaiting and the fabrication of baskets, matting, etc.

Musical Instruments—the preservation, tuning, and repair of pianos,
harmoniums, musical boxes, etc.

Clock and Watch Mending—adapted for intelligent amateurs.

Photography—recent development in rapid processes, handy apparatus,
numerous recipes for sensitizing and developing solutions, and applications to modern illustrative purposes.

London : E. & F. N. SPON, 125, Strand.

New York : 12, Cortlandt Street.

In demy 8vo, cloth, 600 pages, and 1420 Illustrations, 6s.

SPONS'

MECHANICS' OWN BOOK;

A MANUAL FOR HANDICRAFTSMEN AND AMATEURS.

CONTENTS.

Mechanical Drawing—Casting and Founding in Iron, Brass, Bronze, and other Alloys—Forging and Finishing Iron—Sheetmetal Working —Soldering, Brazing, and Burning—Carpentry and Joinery, embracing descriptions of some 400 Woods, over 200 Illustrations of Tools and their uses, Explanations (with Diagrams) of 116 joints and hinges, and Details of Construction of Workshop appliances, rough furniture, Garden and Yard Erections, and House Building—Cabinet-Making and Veneering — Carving and Fretcutting — Upholstery — Painting, Graining, and Marbling — Staining Furniture, Woods, Floors, and Fittings—Gilding, dead and bright, on various grounds—Polishing Marble, Metals, and Wood—Varnishing—Mechanical movements, illustrating contrivances for transmitting motion—Turning in Wood and Metals—Masonry, embracing Stonework, Brickwork, Terracotta, and Concrete—Roofing with Thatch, Tiles, Slates, Felt, Zinc, &c.— Glazing with and without putty, and lead glazing—Plastering and Whitewashing — Paper-hanging — Gas-fitting — Bell-hanging, ordinary and electric Systems — Lighting — Warming — Ventilating — Roads, Pavements, and Bridges — Hedges, Ditches, and Drains — Water Supply and Sanitation—Hints on House Construction suited to new countries.

London: E. & F. N. SPON, 125, Strand.
New York: 12, Cortlandt Street.

www.ingramcontent.com/pod-product-compliance
Lightning Source LLC
Chambersburg PA
CBHW020853020726
47497CB00005B/1391